WHAT FUTURE

THE YEAR'S BEST IDEAS TO RECLAIM, REANIMATE & REINVENT OUR FUTURE

EDITED BY TORIE BOSCH AND ROY SCRANTON

The Unnamed P
Los Angeles, C

The Unnamed Press
P.O. Box 411272
Los Angeles, CA 90041

Published in North America by The Unnamed Press.

ISBN: 978-1-944700-45-4

1 3 5 7 9 10 8 6 4 2

Library of Congress Control Number: 2017914580

This book is distributed by Publishers Group West

Cover design & typeset by Jaya Nicely

CONTENTS

WHAT
FUTURE

INTRODUCTION

We tend to take the future for granted. The human animal is hardwired to believe that tomorrow will follow today in much the same way that today followed yesterday. The sun will rise, birds will sing, and emails will flood our inboxes. Babies will be born, grow into toddlers, enroll in school, graduate from college, get married, and have babies of their own. Labor Day sales will be followed by Black Friday sales will be followed by after-Christmas sales, HBO will roll out a new prestige drama, and Apple or Google or Elon Musk will unveil some new invention to "revolutionize" the way we live. The future, it seems, is inevitable: on track to be much like the present, just with the volume turned up.

Accordingly, how you feel about today anticipates how you feel about tomorrow. If you happen to be American—especially the kind of middle-class or wealthy American who closely follows Silicon Valley's technological advances, who watches with calm pleasure the ever-increasing Dow Jones Industrial Average, and who has been educated in a Whig history of gradual, ineluctable progress—then the future promises to be not only inevitable, but enviable. Science, social justice, capitalism, and democracy will join forces to keep raising the standard of living, bringing people out of poverty, increasing suffrage and tolerance, eliminating war, and eradicating diseases, just like they have for the past two hundred years.

"It may turn out, in the end, that the idea of the future we've come to depend on so much is no longer even available. Or to put it another way: the future is already here, and it's confusing as hell."

Or maybe you have a more pessimistic—you'd say *realistic*—view of things. Maybe you sneer at that pleasing vision of progress as so much utopian pabulum. Maybe you used to feel good about tomorrow, but now? Seriously? Maybe Donald Trump and Peter Thiel and NSA surveillance and the threat of nuclear war and drone strikes and climate change and Russian hacking have got you down. Maybe you see *The Handmaid's Tale* less as a cautionary fable than a threat.

But whether dream or nightmare, the future is as much projection as it is prediction. Think about every vision of future life we've gotten in the past two centuries or so, from Fourier's phalanxes to Marx's proletarian revolution to *Flash Gordon* to *Back to the Future*, even the ones that were surprisingly prophetic, such as *1984* and *Neuromancer*, and you'll realize how much they all missed. As much as any one vision of the future might get wrong, however, thinking of humanity's persistent attempts to imagine tomorrow as a series of win-or-lose propositions misses the point.

Imagining the future isn't completely or even necessarily about forecasting, but about developing concepts and exploring hypothetical scenarios. Imaginary futures, whether they're utopian or dystopian, can serve as places to explore the political and ethical ramifications of our present life, philosophical meditations on contemporary problems, and blueprints for potential pathways. The dystopic vision of *Blade Runner* may not have come to fruition (yet), but its fundamental question—how should humans interact with robot workers?—has gained new currency as more and more robots inhabit our homes and workplaces. Even if Amazon Echo isn't murderous (despite the fact that police *have* tried to use an Echo to solve a murder case), it nevertheless forces us to think about human values. Should children be encouraged to say "please" and "thank you" to Alexa? Might abusing a robot—like abusing a pet—someday be seen as the sign of a troubled mind? Or does treating Alexa as a person encourage a dangerous anthropomorphization that could be leveraged by tech firms hoping to profit from stronger human-robot bonds?

Our visions of the future can mislead us, however, when they lean too heavily on cliché: robot overlords, ethics-free genetic engineering, warp drivers, Mars colonies, free energy, brilliant technologists going rogue and imposing their technologies upon a naive population eager to trade money and privacy for convenience and novelty. (Okay, maybe that last one's more than a cliché.) Too often, our dreams of the future are nothing more than fantasies, dressed up like every Hollywood movie we've ever seen, with cool silvers and blues illustrating a sterile, humanity-lite transcendence, or gritty, desaturated browns and grays coloring a savage descent. Sometimes even the best-intentioned warning can inspire ruinous policy. The infamously awful 31-year-old Computer Fraud and Abuse Act came to fruition after some members of Congress saw the Matthew Broderick film *War Games*; the CFAA was the law under which open-information activist Aaron Swartz was to be prosecuted before he committed suicide in 2013. All too often, our dreams of the future are no more than salesmanship and hokum, infinite happiness for zero down.

Think for a few minutes about how we got to where we are, and that comforting Whig idea of history leading from barbarism to technological utopia starts to look like a con job perpetrated by the 1 percent who have profited most from carbon-fueled capitalism. And once all of our current jobs have been filled by robots, most of us will be freelancers in sharing-economy schemes intended to make the lives of the wealthy as convenient and frictionless as possible. Why pick up your own dry cleaning or juice your own dandelion greens when there's an app for it?

It has always been difficult to distinguish the present from our visions of the future, and the problem is worse now than it has ever been, as the speed of change has accelerated to the point where few (if any) of us have a clear idea of what today actually looks like, much less tomorrow. It may turn out, in the end, that the idea of the future we've come to depend on is no longer even available. To put it another way: the future is already here, and it's confusing as hell.

What Future was inspired by the existential challenge of trying to understand today's tomorrow (or tomorrow's today) as seen through a wobbling global economy, political upheaval, a society distracted and undermined by its technology, and catastrophic climate change. The trend in science fiction for near-future dystopias is no accident—it is becoming alarmingly difficult to imagine or hope or dream beyond our present.

The pieces we've brought together here in *What Future* are some of the best, most interesting, and most prophetic essays and articles we've found about the future: what it might look like, how we might think about it, and what it might mean. We limited our selection to work published in 2016, with a few exceptions, and tried to balance the mix to address some of the most salient issues we see facing human civilization and American culture over the short-to middle-term horizon.

No vision of the future better encapsulates the stirring human ambition to explore new horizons, seek out new ways to live and new forms of civilization, and boldly go where no one has gone before than the dream of throwing ourselves into space and colonizing new worlds. While this dream has, from its beginning, been hopelessly entangled with the bloody history of colonization here on Earth, it has also offered a field in which we could perform our highest ideals, behold new possibilities, and imagine transcending our biological, earthly limits. Our noble, heroic dream of space is, however, for better or for worse, sheer fantasy, as science fiction writer Kim Stanley Robinson patiently shows in "Our Generation Ships Will Sink," from *Boing Boing*. Robinson painstakingly takes up every hope of this age-old dream and demolishes it, reminding us that while our minds might fly through the stars, our bodies have evolved to thrive in a specific ecological niche, and remain dependent not only on oxygen and gravity but on gut bacteria, cycles of day and night, and dependable forms of social existence. "There is no Planet B," Robinson warns. "Earth is our only possible home."

But that doesn't mean that we can't run scientific outposts on nearby planets, much like we've manned space stations in Earth's orbit since Russian cosmonauts boarded *Salyut 1* in 1971. Sheyna Gifford gives us a richly textured personal account of what those outposts might look like in her piece from *Narratively*, "My Life on (Simulated) Mars." Amid describing the daily dramas of rehydrating breakfast, shoveling shit, and coping with spacesuit failure, Gifford finds a lesson in human resilience: "The trick to our continued survival—on Earth and in space—is to build our society so that things can break but...we can then shake it off and keep on running."

Our survival has, so far, been the result of such resilience, part biological evolution and part sociotechnological innovation (from mammoth-bone shelters and the invention of language to nitrogen-fixed fertilizer and the division of labor). Recent technological developments in genetic science have given engineers terrifying new power when it comes to bending evolution to social ends. As Brooke Borel describes in "Genetic Engineering to Clash with Evolution," from *Quanta Magazine*, scientists pairing CRISPR techniques with newly isolated gene drivers will be able to flout the usual rules of natural selection—up to a point. Whether that point comes before or after they do something irrevocable, such as eradicating an entire species of mosquito, for example, remains to be seen.

Some sociotechnological innovations that promise to connect us to others can serve instead to isolate and divide us. Whether we're talking about Twitter bots spreading hatred, Facebook posts inspiring despair, or MMORPG addiction, our embrace of virtual sociality and social media—like our embrace of genetic engineering—remains a dubious choice with ambiguous consequences. In "The Virtual World in a Real Body," originally published in *The Atlantic*, Michael W. Clune questions the popular narrative around VR technology: that it can inspire users to feel empathy for refugees, people with disabilities, and others whose circumstances can be difficult for wealthy, able-bodied Westerners to imagine. "VR immerses your senses in another place," he writes. "But the mind of another person isn't a place, and it can't be entered through your senses alone."

If we can't enter one another's minds through our senses, what about through a computer? In unsettling visions of the future populated by cyborgs, we typically arrive in media res, after upgraded brains have already become commonplace and the regulatory questions have been settled. But in this world, someone has to go first. In a story for *Wired*, Daniel Engber details a real-life neurologist who became his own guinea pig. As part of his lifelong quest to use neural implants to restore speech to people with devastating injuries, Phil Kennedy found himself in Belize, far from the watchful eye of the FDA, undergoing brain surgery. Engber contrasts the high-tech imagined world of cyborgs with the kludgy devices of today, which often appear sadly primitive: "It seems like technology always finds new and better ways to disappoint us, even as it grows more advanced every year."

There are some perennial predictions that we don't want to come true. Elizabeth Kolbert details one in "Our Automated Future," from *The New Yorker*. Kolbert traces the history of machines displacing humans in the workplace, only for technology to create new careers and jobs. It's a comforting vision, until you think about the people who fall into the chasm between the present and the future. "Economic history suggests that this basic pattern will continue, and that the jobs eliminated by Watson and his ilk will be balanced by those created in enterprises yet to be imagined—but not without a good deal of suffering," Kolbert writes.

And who's building these robots? A. M. Gittlitz's *New Inquiry* piece "Let Them Drink Blood" points the finger at Silicon Valley tech bros, describing a future in which rich men maintain their wealth and their bodies by (literally) sucking the life force out of the poor. Gittlitz warns of "a not-too-distant future of cybernetic capitalist reconstruction," in which Peter Thiel (founder of Palantir, killer of *Gawker*, adviser of Trump) and his acolytes found a community of unapologetic selfishness—at sea, in virtual space, and in orbit. Gittlitz compares today's Silicon Valley futurists with the Soviet futurists of the 1920s and '30s, illustrating how the seemingly cutting-edge ideas of the future have long histories.

Silicon Valley's "cybernetic capitalist reconstruction" may be able to depend on robots to do the heavy lifting, but they're still going to need human bodies to staff the riot squads and surveillance teams necessary to quash dissent, at least until they invent Robocop. Federal and local police across the country are already monitoring internet traffic, cracking phones, "employing 'kill switch' technology to cut off live-streaming, using Stingrays to intercept phone calls, [and] flying drones overhead for crowd control," as Malkia Cyril reports in her *Guardian* piece "Black Americans and Encryption." Law enforcement specifically targets black activists in ways that threaten First Amendment rights and give grotesque new life to another bad idea with a long history: white supremacy. Maurice Chammah, in "Policing the Future," published in *The Verge,* takes a look at how police in St. Louis County, Missouri, are using predictive policing software. Chammah asks whether such data-driven policing is more accurate or effective than regular patrols or just layers a veneer of Silicon Valley professionalism over old-school racial profiling.

White nationalism, tech-bro libertarianism, alt-right misogyny, working-class rage, and willful ignorance combined to frightening effect in November 2016 as Donald Trump was elected the 45th president of the United States of America, an election that signified, for Hal Niedzviecki, "the post-political void at the end of the capitalist fantasy." Niedzviecki develops this idea in "Trump Ushers in the Anti-Future Age," from *Dark Mountain,* arguing that while we were busy binge-watching *Stranger Things,* we missed the fact that the spiritual emptiness of a futureless life might inspire some of us to rage. He writes: "Into the vacuum of contemporary politics voided by technological displacement, rapacious capital and the existential threat of climate change arrives Donald Trump, Brexit, Putin, Turkey's Erdoğan and the Philippines' Rodrigo Duterte—anti-future emblems of an unquenchable thirst for simplified belief systems, circles of life, the easy path back to the idyllic pre-literate past."

Meanwhile, as Silicon Valley vampires and Trump's anti-future cabal figure out new ways to prey on human life, take advantage of confused rage, and trick out their lifeboats, the human destruction of the non-human world continues apace in the "sixth extinction," with species across the planet disappearing at a truly astonishing rate. Those species that activists call "charismatic megafauna"—polar bears, tigers, whales—are the poster pups for expanding policy efforts to protect endangered species and their habitats, but no animal better embodies our fraught and confused relationship with the non-human world than the great ape does. In "The Battle for the Great Apes," from *Pacific Standard*, George Johnson reports on efforts to expand legal personhood to great apes by having them recognized as "non-human persons" and "sentient beings."

But sentience may not be all it's cracked up to be. All too often, even when we should see what's coming around the corner, we don't. The sudden death of cash offers an object lesson in this phenomenon. Credit cards, followed by debit cards and now cryptocurrency, have made carrying cash more unnecessary than ever. But it's a transition that happened without much discussion about potential consequences. In Sweden, that conversation is now taking place, as Mallory Pickett describes in her *Wired* article "One Swede Will Kill Cash Forever—Unless His Foe Saves It from Extinction." A case study of what happens when we recognize a sea change only as it washes over us, Pickett's story introduces us to two characters whose furious battle over the future of money is both personal and wide-reaching in its implications.

Now imagine the discussions we'll have when we suddenly realize that we're using robots to care for the elderly. In "The One-Armed Robot That Will Look After Me Until I Die," from *Mosaic Science*, Geoff Watts visits some of the roboticists working to create automatons that can help senior citizens with tasks, monitor their health, and even provide companionship. "Any robot that is going to be able to provide physical help for people is going to have to interact with them on a social level," one expert tells Watts. But if robots can meet the elderly's physical and emotional needs, will senior citizens end up abandoned by other humans? How will *you* think of your robot caregiver? "He"? "It"?

Probably "she," says Janna Avner in her *Real Life* piece "Selfless Devotion." Avner points out that humanoid robots and AIs tend to be portrayed as female: Siri, Alexa, *Her*, Rosie the Robot. That makes a great deal of sense, she writes, because humanity has long leaned on women for caretaking. But it is significant that (mostly male) roboticists default to seeing their subservient creations as female, thereby reinforcing a damaging social dynamic.

Saudi Arabia has taken that dynamic to a further extreme than just about anywhere on Earth. There, each woman has an official guardian—a father, husband, uncle, son—who must give her permission to work, travel, or do just about anything. Famously, the kingdom forbids women to drive. If a male relative cannot drive them, then they must pay a driver, a financial obstacle, even an impossibility, for many women. But what about self-driving cars? Could they create a loophole that might allow women to leapfrog to new freedoms? Not so much, Sarah Aziza says in an article from *Slate*. Aziza, who lived in Saudi Arabia, uses self-driving cars to demonstrate how innovation does not inherently promote social progress.

Technology emerges in relation to preexisting norms. Those norms are themselves shaped by visions of technological progress—or regress—that span a range of political and social commitments. In an article for *The Baffler* titled "Fear of a Feminist Future," Laurie Penny connects the American alt-right's reaction against the feminist liberalism embodied by Hillary Clinton to a perverse fantasy of postapocalyptic neofeudal masculine virtue, while offering several Afrofuturist and feminist science fiction "dystopias" as counterexemplary possibilities for imagining a postpatriarchal future.

It's difficult to imagine a postpatriarchal future that does not include greater gender parity within STEM education and careers, but in "The Disturbing Science Behind Subconscious Gender Bias," an essay published in *The Establishment*, Shoshana Kordova finds STEM careers entangled with patriarchal bias in deep ways. She writes, "Many women...don't even get far enough in the sciences to encounter discrimination in the classroom or the lab. That's because gender bias isn't just about how men view women. Gender stereotypes are at their most insidious when they turn the targets of the stereotypes against themselves." The gender gap in STEM is often called a "pipeline problem," as women drop out, or rather drip out, at various stages of learning or work. Patching up those holes will require more than handing women nail files at STEM recruitment events; it will mean addressing implicit biases that plague all of us from early childhood.

All these issues,, from race and gender to genetic engineering and financial technology, confront us even more urgently as we face together the problem of climate change, perhaps the greatest single challenge to humanity's future. As Bill McKibben argues in "Recalculating the Climate Math," from *The New Republic*, we must transition from a fossil fuel economy *immediately* if we want to have any chance of saving human civilization. Doing so will most likely require not only staying below the 2° Celsius increase in global temperature that leading scientists have agreed is the red line for global warming, but keeping under the 1.5° Celsius increase (which we are currently approaching) suggested by the 2015 Paris Agreement. Despite the ambitious rhetoric in Paris, though, the 2015 accords offered little more than business as usual: international bureaucracies teaming up with nation-states and transnational corporations to shuffle papers, put on a show of resolution, and then go back to letting "the market" solve things. And while the United Nations impotently dithers, the United States under the Trump administration leads the charge to deny and obstruct climate science, shore up unsustainable energy policies, and doom the world to a *Soylent Green* hellscape, the kind of world where the rich harvest their servants' organs and the poor scavenge for old TV parts (not so different from today, in some places...). David Biello imagines just such a world in his essay "The Future Consumed" in order to point up the consequences of our contemporary consumer lifestyle and provoke us into imagining—and living—a different future.

Policy makers in the global south have found some economic and environmental hope in the increasingly cheap power offered by renewable energy. Erica Gies, in an article from *Nature,* asks "Can Wind and Solar Fuel Africa's Future?" She takes us from Morocco to Zambia to South Africa, where governments are working with renewable energy companies to install solar panels and wind turbines and build up a sustainable grid that can solve (or at least avoid) the problems that come with the oil curse. *What Future* editor Roy Scranton explores some of the problems of that curse as it relates to climate change in his essay "Anthropocene City: Houston as Hyperobject," in which he examines how efforts to protect North America's energy capital from a climate-change-fueled superstorm are paralyzed by weak state governance, lack of funds, and a private sector uninterested in protecting the public, forecasting just the kind of disaster that Hurricane Harvey brought down on Houston this August.

The problems climate change presents, however, are not limited to policy. Nature versus culture, U.N. accords, technofixes, democracy, market solutions—none of our old notions seem to have traction when it comes to the "wicked problem" of catastrophic global warming. The need to reconceptualize our relation to a planetary ecosystem in radical transformation is urgent, and in "Hauntings in the Anthropocene," from *Environmental Critique,* novelist Jeff VanderMeer takes up philosopher Timothy Morton's notion of the "hyperobject" to propose that the uncanny weirdness of the Anthropocene might be understood as a kind of haunting—by what has been lost, by what is yet to come, and by that which may remain forever beyond our senses—taking climate change as a kind of poltergeist we can perceive only through its effects.

The future isn't what it used to be, at least since the Industrial Revolution. Poets Laura Riding and Robert Graves were among the first to mark the new quality modernity lent the future, in 1937. Riding and Graves were talking, pessimistically, about the end of the development of human consciousness in the face of totalized technological society: "The future..." they wrote, "contains nothing but scientific development. It is an involuntary spending and manipulation of physical forces, empty of consciousness." While it's easy to scoff at their gloomy view, we ought to remember that the news in 1937 chronicled the ascendance of Nazi Germany and Soviet Russia, the rise of industrialized mass consumer culture, the spread of war in Asia and Europe, and the decline of what had till then been the world's greatest maritime empire. Two years after their pronouncement, the globe would be plunged into a war that would kill more than 60 million people, about 3 percent of the human population. World War II would turn out to be an event unprecedented in human memory, not only for the scale of bloodshed but also for the almost unimaginable technological transformations in global human society it engendered: 1945 is considered the first year of the "great acceleration," a period of staggeringly rapid increases in human population, energy use, water use, urbanization, atmospheric carbon dioxide, number of dams, acres of pavement, global temperature, and several other trends testifying to what has been, according to the International Geosphere-Biosphere Programme, "the most profound transformation of the human relationship with the natural world in the history of humankind."

Eighty years later, it's sometimes hard to imagine what we are more pessimistic about: the present or the future. Consider the stuff that populates our science fiction: Skynet, seasteading, sea level rise, *Mad Max* monster cars, the end of oil, robot warfare, space colonies with even more deeply entrenched inequality, drought, pandemics, extinction. Think about 3D-printed genetic mutant hybrid intersectional transgender artificial intelligence, mass death, and maybe worse, mass immortality. The dystopic certainly makes for a better story; after all, where's the dramatic tension in a peaceful,

"Then think–about how a large portion of organs available for donation come from car collisions. If we choose mass adoption of self-driving vehicles, will we also have to invest more heavily in 3-D printed organs?"

prosperous, egalitarian society? Maybe we're distracting ourselves as we tumble off the cliff into the abyss, or maybe everything will turn out like Elon Musk promises. The future isn't what it used to be, sure, but *plus ça change, plus c'est la même chose*.

For all the change we've lived through and are living through each day, the future seems uncannily like its old self: still a fog, still a dreamscape, still a haunted projection of present-day hopes and fears onto a whirling cloud of technological change we don't understand, political processes we can't seem to predict, and natural dynamics we can't control. The question this anthology's title poses in the form of a declaration—*What Future*—is not merely a question of *which* future, but also a question of what the future itself means as an idea and as a way of conceptualizing our relation to the world. It is also, perhaps, an exclamation of surprise or astonishment. It might also be a grim shrug of disappointment, or even despair. What future, indeed.

The thing is, no matter how grim or gorgeous the future will turn out to be, we have to live there, so we should still try to figure out what it will look like. Perhaps more important, we ought to figure out what we *want* it to look like. We are inclined to talk about "the future" as if it were a predetermined destination, the ending of a movie: once the cold-fusion jet packs show up, the credits will roll and we will have at last arrived where everything is *settled*. But there are endless options for the future, depending on decisions actual people make about work visas, tax incentives, health care, whether robots are acceptable babysitters, whether to use biometrics in lieu of cash or credit cards, whether CRISPR should be used to treat children with significant chromosomal disorders, as an enhancement to ensure certain babies can grow up to be Olympic athletes, or not at all. Each discrete technological advance presents new dilemmas, potential pathways, and ramifications, as every breakthrough emerges entangled with the preexisting world. Seemingly disparate technologies reinforce and complement each other. Think about how self-driving cars promise to save thousands of people each year from traffic accidents caused by human error. Then think—as Ian Adams and Anne Hobson, fellows at the R Street Institute, have discussed—about how a large portion of organs available for donation come from car collisions. If we choose self-driving vehicles, will we also have to invest more heavily in 3D-printed organs?

Choose is the important verb here. Between scientific discovery, gadget invention, and wide-scale adoption lie convoluted processes requiring time, money, policy decisions, and, one hopes, broad discussion. But how do we make those choices? *Who* gets to make them? Should it be the mad geniuses of Silicon Valley and the early adopters who salivate at every new product release? (Remember the suckers who shelled out for Google Glass?) Should it be lawmakers who don't fully understand the technologies they're asked to regulate? We giggle when a legislator refers to the internet as a "series of tubes," but maybe tech policy should be led by the people who use, rather than create, such innovations. Yet whether we follow the lead of Silicon Valley or Washington, D.C., we are following an affluent, highly educated cohort of, to be honest, mostly white dudes. If such a homogenous group is responsible for designing tomorrow, then we will find ourselves in a future that privileges that cohort and reinforces the disparities and structures that created it. To keep that from happening, we need to make sure conversations about the future are fed by a diversity of opinions, backgrounds, and kinds of knowledge. We need to make choices that broaden the conversation and let it go strange places.

While our choices are critical, we also have to recognize that they are constrained and shaped by forces beyond our control. The future isn't simply ours to decide, but is negotiated between human politics, technological development, and the environment we live in. Every new decision and every new innovation produces unintended consequences that we too often don't see until too late. And sometimes we do see the likely consequences but keep making the wrong choice anyway, because humans are self-interested, flawed, and generally short-sighted. Catastrophic climate change—which scientists have been warning us about since at least 1988—is only the most outrageous and dispiriting example of consequences coming down on us from choices intended and unintended. The history of human interactions with the natural world since industrialization—since the dawn of civilization—since there have been humans, really—is a story of dynamic feedbacks within a global ecosystem leading to increasingly severe degradation of the conditions necessary for our species' survival. Part of what it means to think about the future is thinking about those aspects of the future that may well be beyond our control and trying to come to terms with them. Acceptance and adaptation are choices too. Our hope is that the pieces we've included here might help us make better choices—or, at the very least, help us see the choices our future presents a little bit more clearly.

Torie Bosch and Roy Scranton
August 31, 2017

OUR GENERATION SHIPS WILL SINK

Kim Stanley Robinson

(First appeared in Boing Boing, *November 16, 2015)*

"The Earth is the cradle of humanity, but one cannot live in a cradle forever."

—Konstantin Tsiolkovsky

Humanity traveling to the stars is an ancient dream, and a late nineteenth and early twentieth century project, proposed quickly after the first developments in rocketry. The idea spread through world culture, mainly by way of science fiction. Countless stories described people visiting planets orbiting other stars, and by a process of cultural diffusion, space travel became one part of a plausible and widely-held consensus future for humanity, a future we seemed to be moving into with accelerating speed as the twentieth century progressed.

With the enormous successes of *Star Trek* and *Star Wars*, the idea was firmly planted in the popular imagination: if we survived as a species, we would be moving out into the galaxy. This awesome diaspora would mark our maturity or success as a species, and would enable us to outlive the Earth itself, should it suffer a natural disaster or be destroyed by some human folly. The thought of long-term galactic survival for humanity was comforting to some, and in any case it seemed inevitable, humanity's fate or destiny. When we landed people on the moon in 1969, and robots on Mars in 1976, it seemed we were already on the way.

But in the same century the idea spread, we were also learning things that made it seem less and less likely that we could do it. When the notion was first broached, we didn't even know how big the universe was; now we do, and it's bigger than we thought. Meanwhile, the tremendous increase in our knowledge of biology has taught us that human beings are much more complicated than we thought, being in effect complex assemblages interpenetrated with larger ecologies.

These and other findings make a contemporary evaluation of the starfaring plan rather startling: one begins to see it can't be done.

Oh no! For some people this is a disturbing and deeply pessimistic conclusion to come to. Then when you combine that new judgment with the recently discovered problems concerning the plan to terraform and inhabit Mars (presence of perchlorates and absence of nitrogen), you come to an entirely new realization about our species: there is no Planet B.

Earth is our only home.

Oh no again!

This conclusion, startling to some, obvious to others, has ramifications that are worth pondering. If it comes to be a generally agreed on view, it might change how we act as individuals and a civilization. These changes in behavior might turn out to be crucial for our descendants. So although this entire discussion consists of speculations about hypothetical futures, which is to say, science fictions, still they are worth thinking about, as useful orientations in our sense of our own history as a species.

The problems that will keep us from going to the stars can be loosely grouped into categories: physical, biological, ecological, sociological, and psychological. One could add economical, but economic problems are trivial compared to the rest, as economics is amenable to adjustment on demand. Reality is not so tractable.

Physically, the main issue is that the stars are too far away.

This problem has been finessed in many science fiction stories by the introduction of some kind of faster-than-light travel, but really this is not going to happen. It's a convenience employed to get us out into a great story space, a magic carpet that gives us the galaxy. I like that story space very much, but any realistic plan for getting to the stars will require slower-than-light travel, probably quite a bit slower. The usual speed mentioned in these discussions, as keeping a balance between the fastest one can imagine accelerating a spaceship while still being able to decelerate it later, is one-tenth of light speed.

The closest stars are four light-years away, although now we know that this Centauri group has no planets we can terraform. Among other nearby stars, Tau Ceti, twelve light-years away, is now known to have planets in its habitable zone; they are too massive for human inhabitation (five or six g), but they might be orbited by habitable moons. Traveling at one-tenth light speed, a voyage there would take 120 years plus the time needed for acceleration and deceleration, so that people speak of approximately two hundred years transit time.

Thus a crossing to even the closest stars will require a multiple generation effort, and the spaceship will need to be a kind of ark, carrying all the other animals and plants the humans will take with them to their new world. This suggests a very large and complicated machine, which would have to function in the interstellar medium for two centuries or more, with no possibility of resupply, and limited possibilities for repair. The spaceship would also have to contain within it a closed biological life support system, in which all the flows of energy and matter would have to recycle as close to perfectly as possible, minimizing catches or clogs of any kind.

Here is where the biological and ecological problems come to the fore, but sticking for now to purely physical problems, the starship would be exposed to far more radiation than we are on Earth, where the atmosphere and magnetosphere protect us to an extent. Effects of that extra radiation are not fully known, but they won't be good. Cladding would help, but would add to the weight of the ship; the fuel carried for deceleration might serve as cladding en route, but that fuel will get burned as the starship slows down, increasing the starfarers' exposure, already higher than it would have been on Earth.

Lastly, in terms of purely physical problems, if the starship runs into anything substantial (like a couple of kilograms) while moving at a tenth of light speed, the impact could be catastrophic.

These physical problems, especially those concerning propulsion and deceleration, are the ones that have received the most consideration by the starship discussion and advocacy community. As engineering problems they can be given at least hypothetical engineering solutions, using equations from physics that we know to be true. Thus they are, in effect, the easiest problems that starships will face, being relatively straightforward. But they aren't that easy.

Biological problems are harder for humans to solve than physical problems, because biology concerns life, which is extraordinarily complex, and includes emergent properties and other poorly understood behaviors. Ultimately biology is still physics, but it constitutes a more complex set of physical problems, and includes areas we can't explain.

We do know that things go wrong in biological systems, because this happens all the time; living things get sick and die. They also very often eat each other, or exist as diseases for each other. These realities mean that biological and ecological problems are much more intractable than physical problems, and are unsolvable in the enclosed context of a multi-generational starship.

It's a matter of size of community, and its isolation from new inputs. A starship would be something like an island, but an island far more isolated than any island on Earth. Processes identified by island biogeography would apply inside a starship, and many of these processes would be accentuated by the radical isolation. As generations of people, plants, and animals passed, reproductive and evolutionary success would be harmed by genetic bottlenecks, also disease, limits on resources, and so on. The super-islanding effect might cause more species than usual to become smaller, and to mutate in other ways, as one sees on ordinary islands. And because bacteria tend to evolve at faster rates than mammals, complete isolation may lead to the development of a suite of bacteria quite different from what the spaceship was sent off with. All mammals include huge numbers of bacteria living inside them, either symbiotically, parasitically, or without significant interaction, so this more rapid genetic shift in the bacterial community could become a big problem to all the larger creatures. On Earth there is a constant infusion of new bacteria into mammals, which sometimes can lead to bad results, as we know; but overall, it's a necessary aspect of healthy existence.

We are always teamed with many other living creatures. Fifty percent of the DNA in our bodies is not human DNA, and this relatively new discovery is startling, because it forces us to realize that we are not discrete individuals, but biomes, like little forests or swamps. Most of the creatures inside us have to be functioning well for the system as a whole to be healthy. This is a difficult balancing act, and does not work perfectly even on Earth; but divorced from Earth's bacterial load, and thus never able to get infusions of new bacteria, the chances of suffering various immune problems similar to those observed in over-sterile Terran environments will rise markedly.

Because we need a broad array of bacterial companions, one would want to bring along as much of Earth as you could fit into a starship. But even the largest starship would be about one-trillionth the size of Earth, and this necessary miniaturization would almost certainly lead to unknown effects in our bodies.

This leads us to the ecological problems, or perhaps we were there all along, because biology is always ecological, as every living thing is a miniature ecological system. But focusing on the level of the community brings up the problems created by the metabolic flow of substances in a closed biological life support system. These flows, of both living and non-living substances, would have to stay balanced within fairly tight parameters, and they would have to avoid any major rifts or blockages. Cycles of oxygen and carbon dioxide, nitrogen, phosphorus, and many other chemicals and elements would have to occur without major fluxes and without catch-points along the way where the element is getting clogged in the system. Earth experiences large ecological fluxes over time, with build-ups of certain elements (oxygen in the atmosphere, carbon in sedimentary rocks) that force evolutionary processes: whatever is alive has to adapt to the new conditions or go extinct. Both often happen.

These fluxes and build-ups would happen inside a starship too, but as the starfarers would be interested in keeping themselves from going extinct, they would have to manage or finesse all the flows to keep from being harmed by them. This would require supporting almost every other living component of the system, except the diseases they would inevitably carry with them; and if chemicals like phosphorus were bonding to substrates as they cycled in the water cycle, which is something they tend to do, this would be bad for the system as a whole. There would never be a chance for exterior additions to the system, nor any good way to stop the cycles, clean up the substrates, and release clogged chemicals. Nor would it be easy to fight or escape diseases that would have piggybacked their way onto the ship; or to deal with any newly evolved aggressive microbial species suddenly feeding on plants, animals, or humans.

In short, a perfectly recycling ecological system is impossible; Earth is not one, and an isolated system a trillion times smaller than Earth would exacerbate the effects of the losses, build-ups, metabolic rifts, balance swings, clogging, and other actions and reactions. All that could be accomplished by starfarers in such an ark would be to deal with these problems as well as possible, minimizing them so that they might hang on long enough for the starship to reach its destination.

But if they do manage that, their problems would have just begun.

Before discussing the problems caused by arrival at the destination, we should finish sketching the problems during the voyage that could be called sociological and psychological. Here things necessarily get more speculative, but for sure it can be said that the people inside the starship will constitute a small and isolated community compared to the population of Earth. And they will be trapped inside their spaceship, and will have to keep the spaceship functioning in order to survive. So whatever their political organization, whether it be military or anarchic, hierarchical or democratic, the situation itself can be called totalitarian.

By this I mean that their situation will demand certain behaviors to ensure their survival. They will have to tightly control their population; both maximum and minimum human numbers will be necessary, and whatever system they devise to achieve this stability, it will not include individual unconstrained choice. Also, there will be quite a few jobs that will simply have to be filled in order for their life support systems to be maintained. Again, however they manage this issue, people will not be free to do what they want, or to do nothing. So in these areas of reproduction and work, generally regarded as basic to human meaning and political freedom, the society in the starship will have to rigidly control themselves. No matter their methods for achieving this control, they will end up living in some version of a totalitarian state. The spaceship will be their state, and to keep the spaceship functioning, the state will rule.

The psychological effects of all these constraints and problems, including the knowledge that Earth exists light-years away, with a population millions of times bigger than the ship's, and a land surface a trillion times larger, cannot be known for sure. It might very well feel like exile; it might feel like being born and living one's entire life in prison.

Add to this inescapable isolation and confinement the effects of an entire life spent indoors, and it seems likely there would be some bad psychological effects. Indeed it seems like a recipe for psychological disaster, a veritable witch's brew of alienation and resentment. If anyone were to lose their sanity in this situation and decide to escape from it, it might be possible to sabotage the starship itself, destroying it and thus killing everyone aboard. Guarding against such a violent act would be necessary, thus adding to the totalitarian nature of the state, also to its stress and pressure. There would be not just alienation, isolation, and resentment, but also fear.

Of course people are adaptable, and humans tend to take their surroundings for granted. As starship life would be all they had ever known, the starfarers might indeed adapt to their situation. But they would know what that situation was, and know the situation on Earth. They would know that their fate was created for them by ancestors who made the choice to enter the starship, a choice they could never unmake. That might be irritating.

But say all these problems get solved somehow. Say the starship reaches its target star system, and goes into orbit around the planet the starfarers hope to inhabit. What happens then?

The planet or moon they hope to inhabit will be either alive or dead. It will either harbor indigenous life, or it won't. Both possibilities represent terrible problems for the settlers.

There is a third possibility, of course, which is that they won't be able to tell if the planet or moon is alive or not, just as we can't tell now whether Mars is alive or dead. In that case they would still have a problem, they just wouldn't know what kind of problem they had. Finding out could be hard.

If the planet harbors indigenous life, then how that life would interact with Terran life would be impossible to determine without experiment. It might turn out to be no problem, or a small problem, or a fatal problem, but for sure it would have to be investigated before the settlement could safely proceed. If the indigenous life proved to interact badly with Terran life, this would have to be dealt with, if possible. But dealing with it might not be possible. And at what point would people decide that it was safe to come in contact with an alien life form, much less coexist with it over the long haul? That would be a hard call to make.

If the planet turned out to be a dead rock, that would remove the problem of coexisting with an alien, but the planet would then have to be terraformed to make it habitable for Terran life, including humans. This would take many years, possibly centuries, possibly even thousands of years, depending on conditions and resources. Recall that the settlers will only have their single starship to power the effort, and planets or moons with gravity anywhere near Earth's gravity will be large. Terraforming any such body will definitely require a huge application of energy, and thus take a long time. And for most if not all of that time, the settlers will either have to wait in orbit in their starship, where all the starship's problems will still obtain, or they will live in shelters constructed on the surface of the planet, shelters that would be a grounded equivalent of the starship, still harboring most of the problems of a closed biological life support system. Either way, in space or on the new planet, they would still be experiencing most of the problems that the starship gave them during the voyage. Having survived a couple hundred years, could they continue that success much longer?

Hard to say; but for sure, arrival at the destination does not end their problems.

There have been many science fiction stories about starships published, and some have suggested various solutions to the problems outlined above.

One is to send small ships filled with frozen embryos, which would be automatically thawed and birthed on arrival. But this solution ignores the issue of the microbiomes existing inside us; these too would have to be brought along, and even with suites of intestinal bacteria perfectly preserved, calibrated, and introduced into the newborns, there then remains the problem of educating and socializing the new youngsters. Often, if the problem is mentioned at all, the idea seems to be that robots and films and libraries could do the job. Good luck with that!

Another suggestion involves what is often called hibernation, or sometimes cold sleep, or cryonic suspension. In this scenario, an adult population is put into some state of suspended animation, then awakened or reanimated when the ship reaches its destination.

This seems promising at first, and indeed I used the idea myself as an emergency rescue method in my starship novel *Aurora*, so I'm familiar with the suggestion. But if this solution is not to become yet another version of the magic carpet, then it has to be remembered that these suspended passengers would not be completely frozen and inert, for then they would be dead. They are hibernating only; chilled and/or chemically slowed down, but not completely stopped, because we don't know how to restart humans who have completely stopped. The passengers would therefore be living some minimally-active form of life. They would still be alive.

That being the case, they would be aging. Physical problems that they had before being suspended would continue to etiolate; new physical problems might crop up, and proceed slowly along their course. This is what aging means, and slow or fast, it would happen in any living system. It's simply entropy again, rearing its head as it always does.

Because of this unavoidable process, even if we had a very successful method for slowing ourselves down, it would still not stop the passengers from aging and then dying, and that would set a limit on how far they could get. And the distances to the stars are so great that even if the bubble of the area that we could reach were expanded by a hundred times over what a normally living population might reach, that would still represent a small portion of the galaxy. A thousand-light-year trip, taking over ten thousand years, would still only get us out to a bubble representing one percent of the Milky Way. That would include a lot of stars, but how many have just the right planet to fit our needs? And how would we know which ones those might be, in advance of a close examination of them? We would never know where to try to go in the first place, and wouldn't have the luxury of stopping to look around along the way.

So it won't work. But people want to believe in it. And it has to be admitted that all the problems combined together still don't add up to the sheer impossibility of faster-than-light travel. Multi-generational starship travel is simply very, very, very unlikely to succeed. If the odds are something like a million to one, should we try it?

Maybe not.

Should we stop telling the story?

Maybe not. One of the best novels in the history of world literature, Gene Wolfe's *Book of the Long Sun* and *Book of the Short Sun*, a seven-volume saga telling the story of a starship voyage and the inhabitation of a new planetary system, finesses all these problems in ways that allow huge enjoyment of the story it tells. The novel justifies the entertaining of the idea, no doubt about it.

But when we consider how we should behave now, we should keep in mind that the idea that if we wreck Earth we will have somewhere else to go is simply false. That needs to be kept in mind,

"*There is no Planet B! Earth is our only possible home!*

Oh no!

But wait: why is that so bad?"

to set a proper value on our one and only planet, so that a moral hazard is not created that allows us to get sloppy with our caretaking of it.

There is no Planet B! Earth is our only possible home!

Oh no!

But wait: why is that so bad?

Here everyone has to answer for themselves. I'm saying it's not bad at all; it just is, and it can be regarded as a good thing. And good or bad, it just is. That's reality. We are not gods, and anyone who thinks of science as a magic wand, or even as a verb, is making a mistake, a category error sometimes called scientism. Drill down a little harder on these issues, look at the evidence; use the scientific method properly. Limits to what we can do will quickly appear around you.

I'm not saying we shouldn't go into space; we should. We should send people to the moon, and Mars, and the asteroids, and every place we can in the solar system, putting up stations and swapping humans in and out of them. This is not only a beautiful thing to do, but useful in helping us to design a long-term relationship with Earth itself. Space science is an Earth science. The solar system is our neighborhood. But the stars are too far away.

After all that's been said above, I see one possible remaining starship story that could be believed:

Hibernating passengers are sent on a small fast starship to a likely-looking nearby planet, with a load of frozen embryos. Most of the hibernating passengers die en route, but some survive, aging and getting weaker, but alive when the destination is reached.

These ancients proceed to thaw, birth, and raise a cohort of embryos, successfully getting them to the stage of babies and toddlers. But now the hibernators, fully awake and alive, and thus aging at the usual rate, begin to die off. It's a race to get the youngsters raised and educated while there are still any elders alive to do the job. Eventually nine decrepit post-hibernation survivors find themselves caring for seventy-six five-year-olds. Interesting times! This is the heart of the novel.

The planet they landed on luckily seems dead, and has ice on its surface, and even a breathable atmosphere (not likely but not impossible). The elders spread Terran bacteria on the surface, then release all the plants and animals they brought with them, hoping to terraform the place as quickly as possible. The planet has nearly one g, which is a good thing for all Terran creatures' health, but means the planet is about as big as Earth. Terraforming will take a while, perhaps a few centuries.

They all move into a habitat on the surface built by their robots, near a frozen sea. After a couple of decades pass, all the hibernators have died, and the youngsters, all twenty-five years old now, have this new world to inhabit.

Good luck to them! Great story! It could join Joanna Russ's *We Who Are About To...* as one of the truly memorable planetary romances in science fiction. Like that great novel, it would be both interesting and believable—indeed not just believable, but the only starfaring scenario one could possible believe!

If you can.

MY LIFE ON (SIMULATED) MARS

Sheyna Gifford

(First appeared in Narratively, *April 5, 2016)*

Downstairs, a giant, half-British, half-Texan 32-year-old guy in a Hawaiian shirt is listening to death metal and making breakfast burritos. "Tortilla!" his voice emanates from below me in the dome.

Sunday isn't my day to cook breakfast, so I get to stay in bed for a few minutes longer. Still, there's no day off on Mars, or on sMars, as we call the simulated Mars-like environment where I am living for a full year. Strapping a grey pedometer around my wrist, I reach for my iPad, then my electronic badge. One by one, I gather up and put on the other various gadgets and gizmos that track my heart rate, location, and distance to other crewmembers as I go about my life, in this simulated world.

By "simulated" I don't mean to imply that sMars is some kind of virtual-reality game, theme park, or subterranean research facility. It's quite real—as real as my home back in St. Louis. Only instead of a brick-and-mortar two-story on a tree-lined street, it's a 1,200-square-foot dome near the top of Mauna Loa volcano in Hawaii. For the past half year, six simulated astronauts, myself included, have been living, working, eating and conducting experiments on everything from plant growth to virtual reality inside this white geodesic frame as if we were on Mars.

If you have to pick somewhere on Earth to practice for the real thing, this volcano is a pretty good place to start. Outside our small, round windows, the world is a red, rocky wasteland of sorts. Kilometers of old lava, frozen mid-flow, give us a place to practice making measurements and taking samples undisturbed by the sights and sounds of other human life. Apart from the dome, the only easily visible structure is our own solar array. On our side of the portholes, plants grow under glowing lights, giving us the occasional fresh food to eat. The six of us were chosen for our skills, education and availability—Got a year and a lot of degrees? Have we got the job for you!—as well as our willingness to endure 365 days of rehydrated food. Still, we guard these growing green things covetously, as if we are dragons and these tiny tomatoes, sprouts, and blades of grass are our hoard. In that way, and many others, sMars is a pretty good analog for the planet next door.

On our version of Mars, everyone has a call sign. "Ace" is what we call our chief engineer, who has the Sunday morning breakfast shift on sMars. My call sign is "Doc Mom." I didn't choose it, but it's a reasonable fit for a 37-year-old doctor from California who tends to bake cookies and stop fights.

The din of electric guitars gyrates the sleep from my eyes. "Morning, Ace!" I call through the door of my second-floor bunk room.

At this point, I've got every whirring, beating, blinking sensor strapped on properly. I stand up, stretch, and open the door to the second-floor landing. Walking out, I look down and see that the glowing blue-and-red indoor plant lights are already on. I start heading down the twelve stairs towards the kitchen.

Here on sMars, we have full Earth gravity, not the one-third gravity that we'll encounter when we finally get to Mars. We're not really sure what difference that makes on a long-term basis; no one has ever lived in one-third gravity before. But when we're walking around on the hot, barren lava, taking rock samples or looking for a cave big enough to hide in during a radiation storm, we notice the full gravity. Our suits are heavier than they should be. The sun is hotter than it should be (on Mars, it will be one-third as bright). If we fall, we land harder than we would in one-third gravity. It's happened to all of us—to me, most spectacularly, while I was standing completely still. One second my feet were resting on the lid of a channel that used to hold rushing lava. The next, I was standing up to my waist in the channel. As I pulled myself out of the newly formed hole in the ground I thought, "That would have hurt less on Mars."

That was a few months back. These days, I can walk almost normally again. Today, as I come around the corner past the white hydroponics system that sounds vaguely like a toilet that won't stop running, I say "Good morning!" to Tristan, aka Marmot, our space architect, who sits at our dining room table, backlit by the glowing two-foot-wide porthole, eyes wide, curved black headphones in. I can't tell if he's reading email, watching a movie, or designing a spacesuit. Knowing him, it's probably all three.

Without pausing to reply, Marmot puts down his electronic stylus, grabs his white ceramic mug, stands up and says, "Gotta go!" He knocks back the remains of his coffee—an excellent roast made in his home state of Montana and brought up the mountain during one of our periodic resupplies—and scampers off for the airlock.

Just like in the movies, Earth ships us supplies. Because it's a long way up the volcano—not 150 million miles long, but still a haul—we beam lists of the things we need back to mission control. They do their best to pack it all up and shoot it our way via highly-trained humans playing the role of the robots who will do the delivering to real Mars. Not every delivery makes it. When you're on Mars, micrometeorites, solar events and unpredictable atmospheres all stand between you and the nearest grocery store. On sMars, deliveries sometimes meet an untimely end on their way up the volcano. During the last resupply run, we popped the lid on a plastic bin containing the oozing, yellow-belled remains of several tightly-packed jars of ghee (shelf-stable butter). We deal with it. Back

"In space, we don't plan
to *not fail* so much as
we plan to *survive*
failing repeatedly."

on Earth, you can make a stink when the bagger at the supermarket breaks your eggs. Here, we're just happy to have something edible to rehydrate.

Something else you learn to live with on sMars: people sometimes get up and run. They usually have a good reason for it. Earlier this week, seconds before we were supposed to start the five-minute "decompression cycle" that allows us to go outside in our spacesuits, one of my crewmates suddenly bolted out of the airlock. One minute he was standing next to all of us, walking stick in his hand, breath condensing slightly on his faceplate. The next moment, he was gone—back through the airlock doors without an explanation. That's mighty unusual for an astronaut. The three us still in our suits peered uncertainly into the dome. There was our astrobiologist, Cyprien, ripping his suit off. When it was off, he clicked his radio back on, turned to us and said, almost amused, "I smelled burning plastic. My arm was too hot. I think my fan was melting."

See: even in the future, stuff breaks. The trick to our continued survival—on Earth and in space—is to build our society so that things can break but, like a dog that's rushed into a lake after a ball, we can then shake it off and keep on running. When we decide that it matters, it turns out that human beings are actually pretty good at this. For the last fifty years, we've constructed satellites and rovers that survive years, sometimes decades beyond what their original designs projected. The Cassini spacecraft orbiting around Saturn and the Opportunity rover on Mars likely owe their twelve years of success to something called fault tolerance; the ability to accept new courses, unexpected changes, even failures. When Mir was still orbiting the Earth, it was said that any system on the station could fail three times over, and the ship would keep on flying. In space, we don't plan to *not fail* so much as we plan to *survive failing repeatedly.*

Fault tolerance is built into every long-duration space mission, but you can't build it into people. We have to program it into ourselves. Even if we happen to possess the emotional skills before liftoff—patience, open-mindedness to new ideas, foods, sounds, living arrangements—the key to the crew's survival is to remember to run those emotional programs at the appropriate time. To execute the commands that let us live and let live when we're vibrated awake by heavy metal, or when a crewmate runs away the instant we say "good morning." We didn't run after Cyprien, nor would it ever occur to me to go after Marmot now. We're all here trying to survive. Survival means getting along. If someone runs away without pausing to explain why to the rest of the crew, you don't waste a minute wondering if that person is being rude. You assume their space suit is on fire, until proven otherwise.

As Marmot vanishes through the white curtain of the airlock, Ace turns and walks to the pantry to retrieve the Martian version of a chicken: a metal canister full of egg crystals. Just add water and presto: Martian eggs.

I watch Ace reconstitute the egg crystals and add shelf-stable butter to a heavy skillet, and it suddenly occurs to me that on real Mars, they might not have Sundays. That would be a shame. It would be a shame because Sundays on sMars are the days that feel most like a day on Earth. There are no cartoons or newspaper, but there is a bit of a sense on Sundays of being at home with your family.

Saturdays we get up, hop into spacesuits and head out the airlock before ten a.m., or earlier. Weekdays are designated for experiments—NASA's and our own—cooking shifts, raising crops, habitat maintenance, media relations, more Extra-Vehicular Activities (EVAs) and exercise. Monday through Saturday, we get up early, check the weather; grab some food, and get to work growing more food. We fix our equipment, clean our clothes and the dome, make lists of chores that need to get done, and call it a day. If the power is low, we ride our modified electricity-producing bicycle instead of walking on the treadmill. But generally, we do things pretty much the same way that people do back on Earth. In other words, living in simulated space is a lot like living in space would actually be: a farm where the farmhands all have PhDs and an uncanny ability to repair space suits with duct tape.

I make the British-Texan his tea: milk powder and too much sugar. I've never met a doctor from another planet, but I suspect that their lot is a lot like mine. We advise, remind, cajole and even plead sometimes. In the end, the most we can usually do to improve health is continue to suggest viable alternatives. Here is one place where I haven't even bothered, though. First off, I know better than to mess with a British man's tea. Secondly, apart from Sunday mornings, we don't have much around here to remind us of home. Our commander, Carmel, a 27-year-old soil scientist, brought her state flag from Montana and photos of her five-year-old nephew. Ace has a portrait from his wedding day and a miniature model of a Boeing 747 in mid-flight. I brought my brown leather doctor bag. Cheesy? Possibly, but also highly functional. It doesn't just carry all my tools, arranged so that I could find any of them in the dark while half asleep; it also carries my memories. I look at it and can almost see myself walking down a hallway in a hospital on Earth in a white coat. I can see my old self stowing it at the nurses' station while I visited with patients. Here on sMars, it's a luxury, even an indulgence, like Ace's over-sweet tea.

I put his cup on the counter and take five steps to my desk to see what happened at mission control overnight. From my desk, it's a few short steps to the airlock, and just a few more to the biology lab. In space, space is at a premium. That's part of the deal: with being in real space, or simulated space. Your fitness as a potential Martian depends very much on your health, your view on confined spaces, and how much you dig a few critical things: rocks, plants, endless home maintenance projects, and humor. Humor is adaptive to deep space in the same way that running after a mastodon while hurling a spear was adaptive to our ice-age ancestors. By staving off boredom, a serious issue after months of rocks and plants, water restriction, and things nearly catching on fire, humor keeps us going. You can spend two hours disassembling the composting toilet, literally raking half-processed human waste out of a bin with a yellow plastic shovel into garbage bags. It'll be fine, so long as when you finally emerge with the bloated, steaming black sacks slung over your shoulder, as I did that day, you turn to your nearest crew mates and say, "I am the WORST Santa Claus EVER. HO HO HO."

I am a simulated astronaut and I shovel shit, when the job calls for it. I came to space to be a doctor, but more often I end up being a plumber, an electrician, and a mechanic. That's the nature of the game. That's survival. We survive by recycling everything. On Mars, we can't even waste our waste. So what do we do? We shovel, and we patch, and we laugh about it. Laughter is a pure, concentrated form of fault tolerance; a much-needed way out when things go wrong within the tiny spaces of our lives.

On the way back into the kitchen, I stop by the power display and check on our state of charge. That's something else that sets Martians apart: living off the grid 24-seven. To live in space, you have to be down with consulting the power supply before so much as switching on the coffee maker. If you happen to be on a planet with clouds or dust storms, you have to check the weather, too. Power right now doesn't mean power five minutes from now. When you live off the grid on one solar array, we as do, heating, cooling, refrigeration, and, of course, toilet fans take precedence over treadmills, video games, lights, and, yes, even coffee. We've spent nights shivering in our bunks, sometimes, so that we can keep the fans and sample freezers running. At the same time, just try to run a major scientific enterprise without coffee and see how far you get. Fortunately, if it ever came to it, we could grind beans by hand and use the pedicycle to heat water. It's never come to that, but it could, any day now.

Standing on Ace's left, watching him add rehydrated bell peppers and cayenne to the eggs, I'm very grateful that this particular resupply made it safely. On some very primal level, not knowing what's going to die in the next Earth resupply makes you feel the opposite of what most people think about when they think about futuristic space explorers. You can tolerate fault in your Internet provider, your car, boat, blimp, spaceship, life partner, or power system. Wherever you are in the universe, not knowing where your next meal is coming from makes you feel very small, very desperate, and very human.

Leaving the planet doesn't change the basic facts of our human existence. Rather, it highlights the strict limits to our survival. Anywhere in the universe humanity decides to go, we will be packing along a set of near-inviolable biological and psychological requirements. We can leave the planet on which we were born, travel a hundred million miles, and we'll still need what we have always needed: food, water, air, warmth, shelter and company. That last one—company—is as basic a human need as the rest, which is why the cadences of heavy metal crashing up the stairs accompanied by the scent of frying eggs is comforting beyond description.

It's a pretty typical Sunday on sMars. After breakfast, I will set up a virtual-reality experiment, so that the crew can remember what it's like to lie on a beach, or sit in the woods, or stand on a windswept cliff overlooking the sea. It's pretty wild, actually. Your brain knows that you are sitting in a blue beach chair wearing a set of 3-D goggles. Your eyes are telling you that you're gazing over the edge of a 300-foot drop. We're not exactly sure how this is supposed to be relaxing. Maybe it isn't. Maybe it's more about a serious change of scene than anything else. As I wander into the music-filled kitchen, I wonder in passing if anyone will ever sprawl on a beach chair under a pink-tinged heat lamp and watch scenes of Mars, real or simulated: the expanses of red, black and grey lava, some craggy, most worn smooth with time, the magnetic dust clinging to and caking the viewport as the scene goes on.

"Tortilla!?" says the chief engineer. It's half interrogative, half piratical command.

"Yes, please," I reply, looking over our selection of cheeses for the day. Andrzej, the egg-slinging engineer, makes two kinds of cheese that he calls "Phil" and "Geno." They are similar to cream-cheesy brie. The commander Carmel's cheese culture is a white lumpy mix called "Chewbacca." In the end, I reach for my goat cheese culture, a crème fraîche called "Gerard." The music morphs into some medley involving guitars with a lot of reverb and someone angrily screaming, "Up from the bowels of hell he sailed, wielding a tankard of freshly brewed ale. Arrrgh!" On Earth, I would beg the chief engineer to turn that noise off. In simulated space, we do a lot of strange things. We name our cheeses, our bread, and a lot of our plants. We compete for who can take the shortest showers. We debate the best way to build habitats and test theories over dinner. And we learn to listen...to a lot of different things.

"What's this one called?" I shout over the din as a new ballad begins.

"'The Trooper'!" he answers, grinning.

Honestly, it sounded more like fifty ways to bludgeon an amplifier. But the theme certainly fits. "Cool," I say, trying to think of anything else I could possibly say. "Who's it by?"

"Iron Maiden! Sweet, huh?"

"Yeah man," I say, dumping eggs onto my tortilla, smearing it with Gerard, and bopping my head to the almost-existent beat. "I dig it." I pick up my plate and look toward the living room to see if the other Martians care to put down their table saws, virtual-reality goggles, and bacteria long enough to have breakfast.

GENETIC ENGINEERING TO CLASH WITH EVOLUTION

Brooke Borel

(First appeared in Quanta Magazine, *September 8, 2016)*

In a crowded auditorium at New York's Cold Spring Harbor Laboratory in August, Philipp Messer, a population geneticist at Cornell University, took the stage to discuss a powerful and controversial new application for genetic engineering: gene drives.

Gene drives can force a trait through a population, defying the usual rules of inheritance. A specific trait ordinarily has a 50–50 chance of being passed along to the next generation. A gene drive could push that rate to nearly 100 percent. The genetic dominance would then continue in all future generations. You want all the fruit flies in your lab to have light eyes? Engineer a drive for eye color, and soon enough, the fruit flies' offspring will have light eyes, as will their offspring, and so on for all future generations. Gene drives may work in any species that reproduces sexually, and they have the potential to revolutionize disease control, agriculture, conservation and more. Scientists might be able to stop mosquitoes from spreading malaria, for example, or eradicate an invasive species.

The technology represents the first time in history that humans have the ability to engineer the genes of a wild population. As

"On an evolutionary timescale, nothing we do matters. Except, of course, extinction. Evolution doesn't come back from that one."

such, it raises intense ethical and practical concerns, not only from critics but from the very scientists who are working with it.

Messer's presentation highlighted a potential snag for plans to engineer wild ecosystems: Nature usually finds a way around our meddling. Pathogens evolve antibiotic resistance; insects and weeds evolve to thwart pesticides. Mosquitoes and invasive species reprogrammed with gene drives can be expected to adapt as well, especially if the gene drive is harmful to the organism—it'll try to survive by breaking the drive.

"In the long run, even with a gene drive, evolution wins in the end," said Kevin Esvelt, an evolutionary engineer at the Massachusetts Institute of Technology. "On an evolutionary timescale, nothing we do matters. Except, of course, extinction. Evolution doesn't come back from that one."

Gene drives are a young technology, and none have been released into the wild. A handful of laboratory studies show that gene drives work in practice—in fruit flies, mosquitoes and yeast. Most of these experiments have found that the organisms begin to develop evolutionary resistance that should hinder the gene drives. But these proof-of-concept studies follow small populations of organisms. Large populations with more genetic diversity—like the millions of swarms of insects in the wild—pose the most opportunities for resistance to emerge.

It's impossible—and unethical—to test a gene drive in a vast wild population to sort out the kinks. Once a gene drive has been released, there may be no way to take it back. (Some researchers have suggested the possibility of releasing a second gene drive to shut down a rogue one. But that approach is hypothetical, and even if it worked, the ecological damage done in the meantime would remain unchanged.)

The next best option is to build models to approximate how wild populations might respond to the introduction of a gene drive. Messer and other researchers are doing just that. "For us, it was clear that there was this discrepancy—a lot of geneticists have done a great job at trying to build these systems, but they were not concerned that much with what is happening on a population level," Messer said. Instead, he wants to learn "what will happen on the population level, if you set these things free and they can evolve for many generations—that's where resistance comes into play."

At the meeting at Cold Spring Harbor Laboratory, Messer discussed a computer model his team developed, which they described in a paper posted in June on the scientific preprint site biorxiv.org. The work is one of three theoretical papers on gene drive resistance submitted to biorxiv.org in the last five months—the others are from a researcher at the University of Texas, Austin, and a joint team from Harvard University and MIT. (The authors are all working to publish their research through traditional peer-reviewed journals.) According to Messer, his model suggests "resistance will evolve almost inevitably in standard gene drive systems."

It's still unclear where all this interplay between resistance and gene drives will end up. It could be that resistance will render the gene drive impotent. On the one hand, this may mean that releasing the drive was a pointless exercise; on the other hand, some researchers argue, resistance could be an important natural safety feature. Evolution is unpredictable by its very nature, but a handful of biologists are using mathematical models and careful lab experiments to try to understand how this powerful genetic tool will behave when it's set loose in the wild.

Resistance Isn't Futile

Gene drives aren't exclusively a human technology. They occasionally appear in nature. Researchers first thought of harnessing the natural versions of gene drives decades ago, proposing to re-create them with "crude means, like radiation" or chemicals, said Anna Buchman, a postdoctoral researcher in molecular biology at the University of California, Riverside. These genetic oddities, she adds, "could be manipulated to spread genes through a population or suppress a population."

In 2003, Austin Burt, an evolutionary geneticist at Imperial College London, proposed a more finely tuned approach called a homing endonuclease gene drive, which would zero in on a specific section of DNA and alter it.

Burt mentioned the potential problem of resistance—and suggested some solutions—both in his seminal paper and in subsequent work. But for years, it was difficult to engineer a drive in the lab, because the available technology was cumbersome.

With the advent of genetic engineering, Burt's idea became reality. In 2012, scientists unveiled CRISPR, a gene-editing tool that has been described as a molecular word processor. It has given scientists the power to alter genetic information in every organism they have tried it on. CRISPR locates a specific bit of genetic code and then breaks both strands of the DNA at that site, allowing genes to be deleted, added or replaced.

CRISPR provides a relatively easy way to release a gene drive. First, researchers insert a CRISPR-powered gene drive into an organism. When the organism mates, its CRISPR-equipped chromosome cleaves the matching chromosome coming from the other parent. The offspring's genetic machinery then attempts to sew up this cut. When it does, it copies over the relevant section of DNA from the first parent—the section that contains the CRISPR gene drive. In this way, the gene drive duplicates itself so that it ends up on both chromosomes, and this will occur with nearly every one of the original organism's offspring.

Just three years after CRISPR's unveiling, scientists at the University of California, San Diego, used CRISPR to insert inheritable gene drives into the DNA of fruit flies, thus building the system Burt had proposed. Now scientists can order the essential biological tools on the internet and build a working gene drive in mere weeks. "Anyone with some genetics knowledge and a few hundred dollars can do it," Messer said. "That makes it even more important that we really study this technology."

Although there are many different ways gene drives could work in practice, two approaches have garnered the most attention: replacement and suppression. A replacement gene drive alters a specific trait. For example, an anti-malaria gene drive might change a mosquito's genome so that the insect no longer had the ability to pick up the malaria parasite. In this situation, the new genes would quickly spread through a wild population so that none of the mosquitoes could carry the parasite, effectively stopping the spread of the disease.

A suppression gene drive would wipe out an entire population. For example, a gene drive that forced all offspring to be male would make reproduction impossible.

But wild populations may resist gene drives in unpredictable ways. "We know from past experiences that mosquitoes, especially the malaria mosquitoes, have such peculiar biology and behavior," said Flaminia Catteruccia, a molecular entomologist at the Harvard T.H. Chan School of Public Health. "Those mosquitoes are much more resilient than we make them. And engineering them will prove more difficult than we think." In fact, such unpredictability could likely be found in any species.

The three new biorxiv.org papers use different models to try to understand this unpredictability, at least at its simplest level.

The Cornell group used a basic mathematical model to map how evolutionary resistance will emerge in a replacement gene drive. It focuses on how DNA heals itself after CRISPR breaks it (the gene drive pushes a CRISPR construct into each new organism, so it can cut, copy and paste itself again). The DNA repairs itself automatically after a break. Exactly how it does so is determined by chance. One option is called nonhomologous end joining, in which the two ends that were broken get stitched back together in a random way. The result is similar to what you would get if you took a sentence, deleted a phrase, and then replaced it with an arbitrary set of words from the dictionary—you might still have a sentence, but it probably wouldn't make sense. The second option is homology-directed repair, which uses a genetic template to heal the broken DNA. This is like deleting a phrase from a sentence, but then copying a known phrase as a replacement—one that you know will fit the context.

Nonhomologous end joining is a recipe for resistance. Because the CRISPR system is designed to locate a specific stretch of DNA, it won't recognize a section that has the equivalent of a nonsensical word in the middle. The gene drive won't get into the DNA, and it won't get passed on to the next generation. With homology-directed repair, the template could include the gene drive, ensuring that it would carry on.

The Cornell model tested both scenarios. "What we found was it really is dependent on two things: the nonhomologous end-joining rate and the population size," said Robert Unckless, an evolutionary geneticist at the University of Kansas who co-authored the paper as a postdoctoral researcher at Cornell. "If you can't get nonhomologous end joining under control, resistance is inevitable. But resistance could take a while to spread, which means you might be able to achieve whatever goal you want to achieve." For example, if the goal is to create a bubble of disease-proof mosquitoes around a city, the gene drive might do its job before resistance sets in.

The team from Harvard and MIT also looked at nonhomologous end joining, but they took it a step further by suggesting a way around it: by designing a gene drive that targets multiple sites in the same gene. "If any of them cut at their sites, then it'll be fine—the gene drive will copy," said Charleston Noble, a doctoral student at Harvard and the first author of the paper. "You have a lot of chances for it to work."

The gene drive could also target an essential gene, Noble said—one that the organism can't afford to lose. The organism may want to kick out the gene drive, but not at the cost of altering a gene that's essential to life.

The third biorxiv.org paper, from the UT Austin team, took a different approach. It looked at how resistance could emerge at the population level through behavior, rather than within the target sequence of DNA. The target population could simply stop breeding with the engineered individuals, for example, thus stopping the gene drive.

"The math works out that if a population is inbred, at least to some degree, the gene drive isn't going to work out as well as in a random population," said James Bull, the author of the paper and an evolutionary biologist at Austin. "It's not just sequence evolution. There could be all kinds of things going on here, by which populations block [gene drives]," Bull added. "I suspect this is the tip of the iceberg."

Resistance is constrained only by the limits of evolutionary creativity. It could emerge from any spot along the target organism's genome. And it extends to the surrounding environment as well. For example, if a mosquito is engineered to withstand malaria, the parasite itself may grow resistant and mutate into a newly infectious form, Noble said.

Not a Bug, but a Feature?

If the point of a gene drive is to push a desired trait through a population, then resistance would seem to be a bad thing. If a drive stops working before an entire population of mosquitoes is malaria-proof, for example, then the disease will still spread. But at the Cold Spring Harbor Laboratory meeting, Messer suggested the opposite: "Let's embrace resistance. It could provide a valuable safety control mechanism." It's possible that the drive could move just far enough to stop a disease in a particular region, but then stop before it spread to all of the mosquitoes worldwide, carrying with it an unknowable probability of unforeseen environmental ruin.

Not everyone is convinced that this optimistic view is warranted. "It's a false security," said Ethan Bier, a geneticist at the University of California, San Diego. He said that while such a strategy is important to study, he worries that researchers will be fooled into thinking that forms of resistance offer "more of a buffer and safety net than they do."

And while mathematical models are helpful, researchers stress that models can't replace actual experimentation. Ecological systems are just too complicated. "We have no experience engineering systems that are going to evolve outside of our control. We have never done that before," Esvelt said. "So that's why a lot of these modeling studies are important—they can give us a handle on what *might* happen. But I'm also hesitant to rely on modeling and trying to predict in advance when systems are so complicated."

Messer hopes to put his theoretical work into a real-world setting, at least in the lab. He is currently directing a gene drive experiment at Cornell that tracks multiple cages of around 5,000 fruit flies each—more animals than past studies have used to research gene drive resistance. The gene drive is designed to distribute a fluorescent protein through the population. The proteins will glow red under a special light, a visual cue showing how far the drive gets before resistance weeds it out.

Others are also working on resistance experiments: Esvelt and Catteruccia, for example, are working with George Church, a geneticist at Harvard Medical School, to develop a gene drive in mosquitoes that they say will be immune to resistance. They plan to insert multiple drives in the same gene—the strategy suggested by the Harvard / MIT paper.

Such experiments will likely guide the next generation of computer models, to help tailor them more precisely to a large wild population.

"I think it's been interesting because there is this sort of going back and forth between theory and empirical work," Unckless said. "We're still in the early days of it, but hopefully it'll be worthwhile for both sides, and we'll make some informed and ethically correct decisions about what to do."

THE VIRTUAL WORLD IN A REAL BODY

Michael W. Clune

(First appeared in The Atlantic, *April 20, 2016)*

From my first encounter with VR, at an exhibition in Chicago's Navy Pier in the early 1990s, a mist of possibility has surrounded the technology. "This is just a prototype," I was told in various booths and studios over the years. But the era of prototypes has finally ended. With the recent release of a new generation of consumer VR headsets such as the Oculus Rift, you can now move, look around, listen, and pick things up in virtual space. The VR content currently available tends to use this capacity to show you the world through someone else's eyes. You become an astronaut, for example, or a refugee. This is a mistake. The technology can't show you what it's like to inhabit another body. Its true function is to remind you how strange it is to inhabit your own.

VR has lived in the future for so long that the prospect of meeting the real thing made me a little anxious. My initial encounter with it comes in the form of a movie. *Notes on Blindness: Into Darkness* is a 14 minute virtual-reality film that premiered at Sundance in January. It's based on the work of the British writer John Hull, who recorded his experience of going blind. The film's promoters say it lets audiences empathize with the condition of a blind person.

I strap on the headset and find myself in a park. Trees, benches, and joggers shimmer in faint colored outlines. A man with a British accent—Hull, I presume—begins to narrate the sounds he heard in the park, which also happen to be the sounds I hear in my headphones. The click-clack of women in heels, a snatch of conversation. When I turn my head, I can see the walkers coming up the path. I can follow their ghostly outlines as they pass, listen to the sound of their footsteps waxing and then waning as the figures disappear around a bend. The voice describes the sounds I heard.

When I took off the headset, I felt confused. So this is what it's like to be blind, I thought, reading the description of the film again. Does blindness suck out the centers of objects, leaving their outlines intact? If that was true, I felt, I would have heard about it before. And what to make of the bodiless voice describing the sounds I heard? Was I supposed to imagine that John Hull's voice was coming out of my body? Or did it mean that my senses had been smuggled into John Hull's body?

It was definitely cool to be able to turn my head and look around *inside* a film. But I was too busy trying to puzzle out the relation of my experience to the film's ostensible aim—helping me empathize with a blind person—to enjoy it much. I saw several other VR films and I played a couple VR games, including some more visually stimulating than *Notes on Blindness*. But they all left me with a similar sense of disappointment and confusion.

The best and worst thing about virtual space is that you don't need imagination to open it. You put on your headset, and you're there. You require exactly the same amount of imagination to be in the park in *Notes on Blindness,* or in the space station in the VR game *ADR1FT,* as you need to be in the room you currently occupy.

When I put on the headset and turn on *ADR1FT,* I find myself floating in space. A chunk of space debris comes flying at me, I swat it away. I'm running low on oxygen. I see a canister in the distance. I move towards it. The canister gets bigger.

With traditional flat media, there's a gap between what you see and what you experience. For example, I recently got the PC game

FTL. When I load it up, a vaguely cartoonish blueprint of a spaceship appears, along with a tiny cartoonish crew. After about ten minutes, the screen still looks the same, but my experience is utterly different. I'm engaged in an intense space battle. There's an explosion as my ship's oxygen system gets hit by a laser. My crew begins to gasp for air. With the last of my fading power I launch a missile, it misses...

In *FTL*, the primitive flat representation of a spaceship must be opened by my imagination before I can inhabit it. The process is essentially the same for a painting, a novel, or a film. The images in these media *resemble* reality. They are not examples of reality. There's a gap between the marks on the screen, canvas, or page, and the person, spaceship, or park that flickers to life for the audience.

You might think that this gap is a problem, something to be overcome. Lots of people have thought this. What if the artwork didn't consist of a representation of a park? What if it put you in the park itself? In a classic essay, the film critic Andre Bazin calls this "the myth of total cinema," a dream that cinema should progress toward "a total and complete representation of reality."

Now that VR has more or less brought this myth to life, it's clear that the gap between screen and world isn't a flaw, but the source of art's power. Think about empathy. One of the chief pleasures of art is the feeling of being able to see the world through the eyes of another person. Art activates our human capacity to use a visible image to unlock an invisible reality. When the child sees a frowning face, she senses the pain of another. As we watch the shifting expressions of Juliette Binoche's character in *Clouds of Sils Maria*, we feel the mixture of tenderness and jealousy with which the aging actress views her young assistant.

The image is a portal to an unseen world. Its flatness, the two-dimensionality of the shapes in *FTL* or *Clouds of Sils Maria*, forces us to use our imagination to pass through the visible to the invisible. Watching Binoche's flat face, we "see through her eyes." We glimpse something of how her history and desire inflect her vision.

VR interprets the phrase "to see through another's eyes" literally. It places us in a character's visual perspective. This is what VR promotional material means when it speaks of the technology's supposed capacity to enhance "empathy." This month's *Wired* cover story celebrates VR's potential to create a "Wikipedia of experiences." But when the VR film *Clouds over Sidra* puts me in the shoes of a twelve year old girl in the Za'atari camp in Jordan, I don't know what it's like to be her. I only know what it's like to be *me*, seeing the things she sees. But what empathy does—what films, photographs, and literature can do—is to show us the camp as *she* sees it rather than as I do.

VR immerses your senses in another place. But the mind of another person isn't a place, and it can't be entered through your senses alone. Of course, nothing prevents VR from activating your imagination in order to achieve true empathy. Imagine a wanderer in virtual space. He finds a virtual flat screen, or a virtual book. He picks it up, and disappears from the visible world.

When I think of Virtual Reality, I think of films and games, imaginary worlds made real. As I was taking off my headset after my third VR film, a self-described virtual-reality 'freak' I'd chatted with earlier in line was pulling at my sleeve.

"You gotta check this out," he said. He was pointing to a small exhibit in a corner of the large room devoted to VR in the Cleveland International Film Festival. A local video company was showcasing their wares. It turned out they specialized in virtual architectural renderings. Using an Oculus Rift headset, they'd created a VR version of their studio for the festival.

I strapped on the headset, wondering why I'd want to see a local video company's studio. Suddenly, I was in a brightly colored room containing several desks and a video camera. There were some lenses on the desk before me. I turned my head and more of the room scrolled into view. I turned my head back to the desk. I picked up a lens with my hands and held it before my face. A tingling started at the base of my spine.

What's going on? I thought. I've done all this before in VR. Why is this affecting me now?

What astonished me was that I wasn't just *looking*. The technology tied my vision to my body—to the motion of my neck and shoulders. My embodied vision twisted and turned through the space like a snake. The feeling of leaning in towards the desk, seeing the desk move towards me, blew me away. The fine details of its wooden surface magnified as I looked closer. I couldn't believe it. I turn my head and I can see the side of the room? Unreal, I thought. Plus I have hands!

For some time the technician had been gently asking me if I'd seen enough. Now he asked a little less gently.

"That was the most beautiful room I've ever seen," I said quietly, taking off the headset.

Stepping away from the exhibit, the secret of its success hit me. Every other VR experience had been oriented towards some story or task. I was trying to find an oxygen canister for my space suit. Or I was trying to pay attention to the narrator while tracking the sound of a jogger to its source. But here, the whole point was simply that I was in a room. There was nothing to do but to look around, to move my hands and my head. Nothing to distract me from the basic magic of VR: It puts you somewhere else.

It puts *you* somewhere else. Forget about the wonder of seeing the world through someone else's eyes, I thought. Why haven't I realized before how amazing it is to see the world through *my* eyes?

The tingling at the base of my spine wasn't going away. By the time I got outside, it had only gotten stronger. I was having a little trouble walking now. I stopped in the middle of the sidewalk, staring at a parked car. An old black Pontiac. Raindrops shone on its hood. I walked a little closer, I couldn't help it. The raindrops got bigger. Amazing.

I slowly, carefully, turned my head. An entire street scrolled away into the distance. I fell against the car, supporting myself on its hood. Thank God it's a cloudy day, I thought. If the sun was out I don't think I'd be able to handle this.

"Forget about the wonder of seeing the world through someone else's eyes. Why haven't I realized before how amazing it is to see the world through *my* eyes?"

The curious sensation lasted for nearly an hour. I was able to drive home, but I had to drive slowly. I couldn't take the highway. At one point the gray stone of the buildings on either side of the street got so distracting I had to pull over.

I'd accidentally discovered the true function of VR. Being in that virtual studio, moving my head and watching the view change. Leaning in towards the desk and seeing small images get larger. Magic. But when I took off the headset the magic didn't stop. It got stronger. I could still move my head and watch the view change. And now there were many more rooms to explore. Hundreds of cars to look at, thousands of buildings to examine. Millions of drops of rain scintillating in incredible realistic detail on an infinite number of surfaces. I thought of the words of the child in Elizabeth Bishop's poem "In the Waiting Room."

> I knew nothing stranger
> had ever happened, nothing
> stranger could ever happen.

It must have felt like this when I first learned how to walk, I thought, as I drove home. Or even earlier, the first time I raised my infant head from the crib and unpeeled a few more inches of the world.

Right now, I thought. I am experiencing what it's like to have a body right now. I turned my head. I moved my hands on the steering wheel. At present, the highest praise that journalists can lavish on a VR system is to say, as Kevin Kelly does in his *Wired* piece, that "the transition back to the real world...was effortless." To me, this misses the true magic of the technology. VR restored the simple wonder of moving around the world in a body.

THE NEUROLOGIST WHO HACKED HIS BRAIN—AND ALMOST LOST HIS MIND

Daniel Engber

(First appeared in Wired, *January 26, 2016)*

The brain surgery lasted 11 and a half hours, beginning on the afternoon of June 21, 2014, and stretching into the Caribbean predawn of the next day. In the afternoon, after the anesthesia had worn off, the neurosurgeon came in, removed his wire-frame glasses, and held them up for his bandaged patient to examine. "What are these called?" he asked.

Phil Kennedy stared at the glasses for a moment. Then his gaze drifted up to the ceiling and over to the television. "Uh…uh…ai…aiee," he stammered after a while, "…aiee…aiee…aiee."

"It's OK, take your time," said the surgeon, Joel Cervantes, doing his best to appear calm. Again Kennedy attempted to respond. It looked as if he was trying to force his brain to work, like someone with a sore throat who bears down to swallow.

Meanwhile, the surgeon's mind kept circling back to the same uneasy thought: "I shouldn't have done this."

When Kennedy had arrived at the airport in Belize City a few days earlier, he had been lucid and precise, a 66-year-old with the stiff, authoritative good looks of a TV doctor. There had been nothing wrong with him, no medical need for Cervantes to open his skull. But Kennedy wanted brain surgery, and he was willing to pay $30,000 to have it done.

Kennedy was himself once a famous neurologist. In the late 1990s he made global headlines for implanting several wire electrodes in the brain of a paralyzed man and then teaching the locked-in patient to control a computer cursor with his mind. Kennedy called his patient the world's "first cyborg," and the press hailed his feat as the first time a person had ever communicated through a brain-computer interface. From then on, Kennedy dedicated his life to the dream of building more and better cyborgs and developing a way to fully digitize a person's thoughts.

Now it was the summer of 2014, and Kennedy had decided that the only way to advance his project was to make it personal. For his next breakthrough, he would tap into a healthy human brain. His own.

Hence Kennedy's trip to Belize for surgery. A local orange farmer and former nightclub owner, Paul Powton, had managed the logistics of Kennedy's operation, and Cervantes—Belize's first native-born neurosurgeon—wielded the scalpel. Powton and Cervantes were the founders of Quality of Life Surgery, a medical tourism clinic that treats chronic pain and spinal disorders and also specializes these days in tummy tucks, nose jobs, manboob reductions, and other medical enhancements.

At first the procedure that Kennedy hired Cervantes to perform—the implantation of a set of glass-and-gold-wire electrodes beneath the surface of his own brain—seemed to go quite well. There wasn't much bleeding during the surgery. But his recovery was fraught with problems. Two days in, Kennedy was sitting on his bed when, all of a sudden, his jaw began to grind and chatter, and one of his hands began to shake. Powton worried that the seizure would break Kennedy's teeth.

His language problems persisted as well. "He wasn't making sense anymore," Powton says. "He kept apologizing, 'Sorry, sorry,' because he couldn't say anything else." Kennedy could still utter syllables and a few scattered words, but he seemed to have lost the glue that bound them into phrases and sentences. When Kennedy grabbed a pen and tried to write a message, it came out as random letters scrawled on a page.

At first Powton had been impressed by what he called Kennedy's Indiana Jones approach to science: tromping off to Belize, breaking the standard rules of research, gambling with his own mind. Yet now here he was, apparently locked in. "I thought we had damaged him for life," Powton says. "I was like, what have we done?"

Of course, the Irish-born American doctor knew the risks far better than Powton and Cervantes did. After all, Kennedy had invented those glass-and-gold electrodes and overseen their implantation in almost a half dozen other people. So the question wasn't what Powton and Cervantes had done to Kennedy—but what Phil Kennedy had done to himself.

For about as long as there have been computers, there have been people trying to figure out a way to control them with our minds. In 1963 a scientist at Oxford University reported that he had figured out how to use human brain waves to control a simple slide projector. Around the same time, a Spanish neuroscientist at Yale University, José Delgado, grabbed headlines with a grand demonstration at a bullring in Córdoba, Spain. Delgado had invented a device he called a stimoceiver—a radio-controlled brain implant that could pick up neural signals and deliver tiny shocks to the cortex. When Delgado stepped into the ring, he flashed a red cape to incite the bull to charge. As the animal drew close, Delgado pressed two buttons on his radio transmitter: The first triggered the bull's caudate nucleus and slowed the animal to a halt; the second made it turn and trot off toward a wall.

Delgado dreamed of using his electrodes to tap directly into human thoughts: to read them, edit them, improve them. "The human race is at an evolutionary turning point. We're very close to having the power to construct our own mental functions," he told *The New York Times* in 1970, after trying out his implants on mentally ill human subjects. "The question is, what sort of humans would we like, ideally, to construct?"

Not surprisingly, Delgado's work made a lot of people nervous. And in the years that followed, his program faded, beset by controversy, starved of research funding, and stymied by the complexities of the brain, which was not as susceptible to simple hot-wiring as Delgado had imagined.

In the meantime, scientists with more modest agendas—who wanted simply to decipher the brain's signals, rather than to grab civilization by the neurons—continued putting wires in the heads of laboratory animals. By the 1980s neuroscientists had figured out that if you use an implant to record signals from groups of cells in, say, the motor cortex of a monkey, and then you average all their firings together, you can figure out where the monkey means to move its limb—a finding many regarded as the first major step toward developing brain-controlled prostheses for human patients.

But the traditional brain electrode implants used in much of this research had a major drawback: The signals they picked up were notoriously unstable. Because the brain is a jellylike medium, cells sometimes drift out of range while they're being recorded or end up dying from the trauma of colliding with a pointy piece of metal. Eventually electrodes can get so caked with scar tissue that their signals fade completely.

Phil Kennedy's breakthrough—the one that would define his career in neuroscience and ultimately set him on a path to an operating table in Belize—started out as a way to solve this basic bioengineering problem. His idea was to pull the brain inside the electrode so the electrode would stay safely anchored inside the brain. To do this, he affixed the tips of some Teflon-coated gold wires inside a hollow glass cone. In the same tiny space, he inserted another crucial component: a thin slice of sciatic nerve. This crumb of biomaterial would serve to fertilize the nearby neural tissue, enticing microscopic arms from local cells to unfurl into the cone. Instead of plunging a naked wire into the cortex, Kennedy would coax nerve cells to weave their tendriled growths around the implant, locking it in place like a trellis ensnarled in ivy. (For human subjects he would replace the sciatic nerve with a chemical cocktail known to stimulate neural growth.)

The glass cone design seemed to offer an incredible benefit. Now researchers could leave their wires in situ for long stretches of time. Instead of catching snippets of the brain's activity during single sessions in the lab, they could tune in to lifelong soundtracks of the brain's electrical chatter.

Kennedy called his invention the neurotrophic electrode. Soon after he came up with it, he quit his academic post at Georgia Tech and started up a biotech company called Neural Signals. In 1996, after years of animal testing, Neural Signals received approval from the FDA to implant Kennedy's cone electrodes in human patients, as a possible lifeline for people who had no other way to move or speak. And in 1998, Kennedy and his medical collaborator, Emory University neurosurgeon Roy Bakay, took on the patient who would make them scientific celebrities.

Johnny Ray was a 52-year-old drywall contractor and Vietnam veteran who had suffered a stroke at the base of his brain. The injury had left him on a ventilator, stuck in bed, and paralyzed except for slight twitchings of his face and shoulder. He could answer simple questions by blinking twice for "yes" and once for "no."

Since Ray's brain had no way to pass its signals down into his muscles, Kennedy tried to wiretap Ray's head to help him communicate. Kennedy and Bakay placed electrodes in Ray's primary motor cortex, the patch of tissue that controls basic voluntary movements. (They found the perfect spot by first putting Ray into an MRI machine and asking him to imagine moving his hand. Then they put the implant on the spot that lit up most brightly in his fMRI scans.) Once the cones were in place, Kennedy hooked them up to a radio transmitter implanted on top of Ray's skull, just beneath the scalp.

Three times a week, Kennedy worked with Ray, trying to decode the waves from his motor cortex and then turn them into actions. As time went by, Ray learned to modulate the signals from his implant just by thinking. When Kennedy hooked him up to a computer, he was able to use those modulations to control a cursor on the screen (albeit only along a line from left to right). Then he'd twitch his shoulder to trigger a mouseclick. With this setup, Ray could pick out letters from an onscreen keyboard and very slowly spell out words.

"This is right on the cutting edge, it's Star Wars stuff," Bakay told an audience of fellow neurosurgeons in October 1998. A few weeks later, Kennedy presented their results at the annual conference of the Society for Neuroscience. That was enough to send the Amazing Story of Johnny Ray—once locked in, now *typing with his mind*—into newspapers all around the country and the world. That December both Bakay and Kennedy were guests on *Good Morning America*. In January 1999, news of their experiment appeared in *The Washington Post*. "As Philip R. Kennedy, physician and inventor, prepares a paralyzed man to operate a computer with his thoughts," the article began, "it briefly seems possible a historic scene is unfolding in this hospital room and that Kennedy might be a new Alexander Graham Bell."

In the aftermath of his success with Johnny Ray, Kennedy seemed to be on the verge of something big. But when he and Bakay put brain implants in two more locked-in patients in 1999 and 2002, their cases didn't push the project forward. (One patient's incision didn't close and the implant had to be removed; the other patient's disease progressed so rapidly as to make Kennedy's neural recordings useless.) Ray himself died from a brain aneurysm in the fall of 2002.

Meanwhile, other labs were making progress with brain-controlled prostheses, but they were using different equipment—usually small tabs, measuring a couple of millimeters square, with dozens of naked wire leads protruding down into the brain. In the format wars of the tiny neural-implants field, Kennedy's glass-and-cone electrodes were looking more and more like Betamax: a viable, promising technology that ultimately didn't take hold.

It wasn't just hardware that set Kennedy apart from the other scientists working on brain-computer interfaces. Most of his colleagues were focused on a single type of neurally controlled prosthesis, the kind the Pentagon liked to fund through Darpa: an implant that would help a patient (or a wounded veteran) use prosthetic limbs. By 2003 a lab at Arizona State University had put a set of implants inside a monkey that allowed the animal to bring a piece of orange to its mouth with a mind-controlled robotic arm. Some years later researchers at Brown University reported that two paralyzed patients had learned to use implants to control robot arms with such precision that one could take a swig of coffee from a bottle.

But Kennedy was less interested in robot arms than in human voices. Ray's mental cursor showed that locked-in patients could share their thoughts through a computer, even if those thoughts did dribble out like tar pitch at three characters per minute. What if Kennedy could build a brain-computer interface that flowed as smoothly as a healthy person's speech?

In many ways, Kennedy had taken on the far greater challenge. Human speech is immensely more complicated than any movement of a limb. What seems to us a basic action—formulating words—requires the coordinated contraction and release of more than 100 different muscles, ranging from the diaphragm to those of the tongue and lips. To build a working speech prosthesis of the kind Kennedy imagined, a scientist would have to figure out a way to read all the elaborate orchestration of vocal language from the output of a handful of electrodes.

So Kennedy tried something new in 2004, when he put his implants in the brain of one last locked-in patient, a young man named Erik Ramsey, who had been in a car accident and suffered a brain stem stroke like Johnny Ray's. This time Kennedy and Bakay did not place the cone electrodes in the part of the motor cortex that controls the arms and hands. They pushed the wires farther down a strip of brain tissue that drapes along the sides of the cerebrum like a headband. At the bottom of this region lies a patch of neurons that sends signals to the muscles of the lips and jaw and tongue and larynx. That's where Ramsey got his implant, 6 millimeters deep.

Using this device, Kennedy taught Ramsey to produce simple vowel sounds through a synthesizer. But Kennedy had no way of knowing how Ramsey really felt or what exactly was going on in his head. Ramsey could respond to yes-no questions by moving his eyes up or down, but this method faltered because Ramsey had eye problems. Nor was there any way for Kennedy to corroborate his language trials. He'd asked Ramsey to imagine words while he recorded signals from Ramsey's brain—but of course Kennedy had no way of knowing whether Ramsey really "said" the words in silence.

Ramsey's health declined, as did the electronics for the implant in his head. As the years went by, Kennedy's research program suffered too: His grants were not renewed; he had to let his engineers and lab techs go; his partner, Bakay, died. Now Kennedy worked alone or with temporary hired help. (He still spent business hours treating patients at his neurology clinic.) He felt sure he would make another breakthrough if he could just find another patient—ideally someone who could speak out loud, at least at first. By testing his implant on, say, someone in the early stages of a neurodegenerative disease like ALS, he'd have the chance to record from neurons while the person talked. That way, he could figure out the correspondence between each specific sound and neural cue. He'd have the time to train his speech prosthesis—to refine its algorithm for decoding brain activity.

But before Kennedy could find his ALS patient, the FDA revoked its approval for his implants. Under new rules, unless Kennedy could demonstrate that they were safe and sterile—a requirement that would itself require funding that he didn't have—he says he was banned from using his electrodes on any more human subjects.

But Kennedy's ambition didn't dim; if anything, it overflowed. In the fall of 2012, he self-published a science fiction novel called *2051*, which told the story of Alpha, an Irish-born neural electrode pioneer like Kennedy who lived, at the age of 107, as the champion and exemplar of his own technology: a brain wired up inside a 2-foot-tall life-support robot. The novel provided a kind of outline for Kennedy's dreams: His electrodes wouldn't simply be a tool for helping locked-in patients to communicate but would also be the engine of an enhanced and cybernetic future in which people live as minds in metal shells.

By the time he published his novel, Kennedy knew what his next move would be. The man who had become famous for implanting the very first brain-machine communication interface inside a human patient would once again do something that had never been done before. He had no other choices left. "What the hell," he thought. "I'll just do it on myself."

A few days after the operation in Belize, Powton paid one of his daily visits to the guesthouse where Kennedy was convalescing, a bright white villa a block away from the Caribbean. Kennedy's recovery had continued to go poorly: The more effort he put into talking, the more he seemed to get locked up. And no one from the US, it became clear, was coming to take the doctor off Powton and Cervantes' hands. When Powton called Kennedy's fiancée and told her about the complications, she didn't express much sympathy. "I tried stopping him, but he wouldn't listen," she said.

On this particular visit, though, things started to look up. It was a hot day, and Powton brought Kennedy a lime juice. When the two men went out into the garden, Kennedy tilted back his head and let out an easy and contented sigh. "It feels good," he blurted after taking a sip.

After that, Kennedy still had trouble finding words for things—he might look at a pencil and call it a pen—but his fluency improved. Once Cervantes felt his client had gotten halfway back to normal, he cleared him to go home. His early fears of having damaged Kennedy for life turned out to be unfounded; the language loss that left his patient briefly locked in was just a symptom of postoperative brain swelling. With that under control, he would be fine.

By the time Kennedy was back at his office seeing patients just a few days later, the clearest remaining indications of his Central American adventure were some lingering pronunciation problems and the sight of his shaved and bandaged head, which he sometimes hid beneath a multicolored Belizean hat. For the next several months, Kennedy stayed on anti-seizure medications as he waited for his neurons to grow inside the three cone electrodes in his skull.

Then, in October that same year, Kennedy flew back to Belize for a second surgery, this time to have a power coil and radio transceiver connected to the wires protruding from his brain. That surgery went fine, though both Powton and Cervantes were nonplussed at the components that Kennedy wanted tucked under his scalp. "I was a little surprised they were so big," Powton says. The electronics had a clunky, retro look to them. Powton, who tinkers with drones in his spare time, was mystified that anyone would sew such an old-fangled gizmo inside his head: "I was like, 'Haven't you heard of microelectronics, dude?'"

Kennedy began the data-gathering phase of his grand self-experiment as soon as he returned home from Belize for the second time. The week before Thanksgiving, he went into his lab and balanced a magnetic power coil and receiver on his head. Then he started to record his brain activity as he said different phrases out loud and to himself—things like "I think she finds the zoo fun" and "The joy of a job makes a boy say wow"—while tapping a button to help sync his words with his neural traces, much like the way a filmmaker's clapper board syncs picture and sound.

Over the next seven weeks, he spent most days seeing patients from 8 am until 3:30 pm and then used the evenings after work to run through his self-administered battery of tests. In his laboratory notes he is listed as Subject PK, as if to anonymize himself. His notes show that he went into the lab on Thanksgiving and on Christmas Eve.

The experiment didn't last as long as he would have liked. The incision in his scalp never fully closed over the bulky mound of his electronics. After having had the full implant in his head for a total of just 88 days, Kennedy went back under the knife. But this time he didn't bother going to Belize: A surgery to safeguard his health needed no approval from the FDA and would be covered by his regular insurance.

On January 13, 2015, a local surgeon opened up Kennedy's scalp, snipped the wires coming from his brain, and removed the power coil and transceiver. He didn't try to dig around in Kennedy's cortex for the tips of the three glass cone electrodes that were embedded there. It was safer to leave those where they lay, enmeshed in Kennedy's brain tissue, for the rest of his life.

Kennedy's lab sits in a leafy office park on the outskirts of Atlanta, in a yellow clapboard house. A shingle hanging out front identifies Suite B as the home of the Neural Signals Lab. When I meet Kennedy there one day in May 2015, he's dressed in a tweed jacket and a blue-flecked tie, and his hair is neatly parted and brushed back from his forehead in a way that reveals a small depression in his left temple. "That's when he was putting the electronics in," Kennedy says with a slight Irish accent. "The retractor pulled on a branch of the nerve that went to my temporalis muscle. I can't lift this eyebrow." Indeed, I notice that the operation has left his handsome face with an asymmetric droop.

Kennedy agrees to show me the video of his first surgery in Belize, which has been saved to an old-fashioned CD-ROM. As I mentally prepare myself to see the exposed brain of the man standing next to me, Kennedy places the disc into the drive of a desktop computer running Windows 95. It responds with an awful grinding noise, like someone slowly sharpening a knife.

The disc takes a long time to load—so long that we have time to launch into a conversation about his highly unconventional research plan. "Scientists have to be individuals," he says. "You can't do science by committee." As he goes on to talk about how the US too was built by individuals and not committees, the disc drive's grunting takes on the timbre of a wagon rolling down a rocky trail: *ga-chugga-chug, ga-chugga-chug*. "Come on, machine!" he says, interrupting his train of thought as he clicks impatiently at some icons on the screen. "Oh for heaven's sake, I just *have* inserted the disc!"

"I think people overrate brain surgery as being so terribly dangerous," he goes on. "Brain surgery is not that difficult." *Ga-chugga-chug, ga-chugga-chug, ga-chugga-chug*. "If you've got something to do scientifically, you just have to go and do it and not listen to naysayers."

At last a video player window opens on the PC, revealing an image of Kennedy's skull, his scalp pulled away from it with clamps. The grunting of the disc drive is replaced by the eerie, squeaky sound of metal bit on bone. "Oh, so they're still drilling my poor head," he says as we watch his craniotomy begin to play out onscreen.

"Just helping ALS patients and locked-in patients is one thing, but that's not where we stop," Kennedy says, moving on to the big picture. "The first goal is to get the speech restored. The second goal is to restore movement, and a lot of people are working on that—that'll happen, they just need better electrodes. And the third goal would then be to start enhancing normal humans."

He clicks the video ahead, to another clip in which we see his brain exposed—a glistening patch of tissue with blood vessels crawling all along the top. Cervantes pokes an electrode down into Kennedy's neural jelly and starts tugging at the wire. Every so often a blue-gloved hand pauses to dab the cortex with a Gelfoam to stanch a plume of blood.

"Your brain will be infinitely more powerful than the brains we have now," Kennedy continues, as his brain pulsates onscreen. "We're going to extract our brains and connect them to small computers that will do everything for us, and the brains will live on."

"You're excited for that to happen?" I ask.

"Pshaw, yeah, oh my God," he says. "This is how we're evolving."

Sitting there in Kennedy's office, staring at his old computer monitor, I'm not so sure I agree. It seems like technology always finds new and better ways to disappoint us, even as it grows more advanced every year. My smartphone can build words and sentences from my sloppy finger-swipes. But I still curse at its mistakes. (Damn you, autocorrect!) I know that, around the corner, technology far better than Kennedy's juddering computer, his clunky electronics, and my Google Nexus 5 phone is on its way. But will people really want to entrust their brains to it?

On the screen, Cervantes jabs another wire through Kennedy's cortex. "The surgeon is very good, actually, a very nice pair of hands," Kennedy said when we first started watching the video. But now he deviates from our discussion about evolution to bark orders at the screen, like a sports fan in front of a TV. "No, don't do that, don't lift it up," Kennedy says to the pair of hands operating on his brain. "It shouldn't go in at that angle," he explains to me before turning back to the computer. "Push it in more than that!" he says. "OK, that's plenty, that's plenty. Don't push anymore!"

These days, invasive brain implants have been going out of style. The major funders of neural prosthesis research favor an approach that involves laying a flat grid of electrodes, 8 by 8 or 16 by 16 of them, across the naked surface of the brain. This method, called electrocorticography, or ECoG, provides a more blurred-out, impressionistic measure of activity than Kennedy's: Instead of tuning to the voices of single neurons, it listens to a bigger chorus—or, I suppose, committee—of them, as many as hundreds of thousands of neurons at a time.

Proponents of ECoG argue that these choral traces can convey enough information for a computer to decode the brain's intent—even what words or syllables a person means to say. Some smearing of the data might even be a boon: You don't want to fixate on a single wonky violinist when it takes a symphony of neurons to move your vocal cords and lips and tongue. The ECoG grid can also safely stay in place under the skull for a long time, perhaps even longer than Kennedy's cone electrodes. "We don't really know what the limits are, but it's definitely years or decades," says Edward Chang, a surgeon and neurophysiologist at UC San Francisco, who has become one of the leading figures in the field and who is working on a speech prosthesis of his own.

Last summer, as Kennedy was gathering his data to present it at the 2015 meeting of the Society for Neuroscience, another lab published a new procedure for using computers and cranial implants to decode human speech. Called Brain-to-Text, it was developed at the Wadsworth Center in New York in collaboration with researchers in Germany and the Albany Medical Center, and it was tested on seven epileptic patients with implanted ECoG grids. Each subject was asked to read aloud—sections of the Gettysburg Address, the story of Humpty Dumpty, John F. Kennedy's inaugural, and an anonymous piece of fan fiction related to the TV show *Charmed*—while their neural data was recorded. Then the researchers used the ECoG traces to train software for converting neural data into speech sounds and fed its output into a predictive language model—a piece of software that works a bit like the speech-to-text engine on your phone—that could guess which words were coming based on what had come before.

Incredibly, the system kind of worked. The computer spat out snippets of text that bore more than a passing resemblance to Humpty Dumpty, *Charmed* fan fiction, and the rest. "We got a relationship," says Gerwin Schalk, an ECoG expert and coauthor of the study. "We showed that it reconstructed spoken text much better than chance." Earlier speech prosthesis work had shown that individual vowel sounds and consonants could be decoded from the brain; now Schalk's group had shown that it's possible—though difficult and error-prone—to go from brain activity to fully spoken sentences.

But even Schalk admits that this was, at best, a proof of concept. It will be a long time before anyone starts sending fully formed thoughts to a computer, he says—and even longer before anyone finds it really useful. Think about speech-recognition software, which has been around for decades, Schalk says. "It was probably 80 percent accurate in 1980 or something, and 80 percent is a pretty remarkable achievement in terms of engineering. But it's useless in the real world," he says. "I *still* don't use Siri, because it's not good enough."

In the meantime, there are far simpler and more functional ways to help people who have trouble speaking. If a patient can move a finger, he can type out messages in Morse code. If a patient can move her eyes, she can use eye-tracking software on a smartphone. "These devices are dirt cheap," Schalk says. "Now you want to replace one of these with a $100,000 brain implant and get something that's a little better than chance?"

I try to square this idea with all the stunning cyborg demonstrations that have made their way into the media over the years—people drinking coffee with robotic arms, people getting brain implants in Belize. The future always seems so near at hand, just as it did a half century ago when José Delgado stepped into that bullring. One day soon we'll all be brains inside computers; one day soon our thoughts and feelings will be uploaded to the Internet; one day soon our mental states will be shared and data-mined. We can already see the outlines of this scary and amazing place just on the horizon—but the closer we get, the more it seems to fall back into the distance.

Kennedy, for one, has grown tired of this Zeno's paradox of human progress; he has no patience for always getting halfway to the future. That's why he adamantly pushes forward: to prepare us all for the world he wrote about in *2051*, the one that Delgado believed was just around the corner.

When Kennedy finally did present the data that he'd gathered from himself—first at an Emory University symposium last May and then at the Society for Neuroscience conference in October—some of his colleagues were tentatively supportive. By taking on the risk himself, by working alone and out-of-pocket, Kennedy managed to create a sui generis record of language in the brain, Chang says: "It's a very precious set of data, whether or not it will ultimately hold the secret for a speech prosthetic. It's truly an extraordinary event." Other colleagues found the story thrilling, even if they were somewhat baffled: In a field that is constantly hitting up against ethical roadblocks, this man they'd known for years, and always liked, had made a bold and unexpected bid to force brain research to its destiny. Still other scientists were simply aghast. "Some thought I was brave, some thought I was crazy," Kennedy says.

In Georgia, I ask Kennedy if he'd ever do the experiment again. "On myself?" he says. "No. I shouldn't do this again. I mean, certainly not on the same side." He taps his temple, where the cone electrode tips are still lodged. Then, as if energized by the idea of putting implants on the other side of his brain, he launches into plans for making new electrodes and more sophisticated implants; for getting back the FDA's approval for his work; for finding grants so that he can pay for everything.

"No, I shouldn't do the other side," he says finally. "Anyway, I don't have the electronics for it. Ask me again when we've built them." Here's what I take from my time with Kennedy, and from his garbled answer: You can't always plan your path into the future. Sometimes you have to build it first.

OUR AUTOMATED FUTURE

Elizabeth Kolbert

(First appeared in The New Yorker, *December 19 & 26, 2016)*

There are many accounts of the genesis of Watson. The most popular, which is not necessarily the most accurate—and this is the sort of problem that Watson himself often stumbled on—begins in 2004, at a steakhouse near Poughkeepsie. One evening, an I.B.M. executive named Charles Lickel was having dinner there when he noticed that the tables around him had suddenly emptied out. Instead of finishing their sirloins, his fellow-diners had rushed to the bar to watch "Jeopardy!" This was deep into Ken Jennings's seventy-four-game winning streak, and the crowd around the TV was rapt. Not long afterward, Lickel attended a brainstorming session in which participants were asked to come up with I.B.M.'s next "grand challenge." The firm, he suggested, should take on Jennings.

I.B.M. had already fulfilled a similar "grand challenge" seven years earlier, with Deep Blue. The machine had bested Garry Kasparov, then the reigning world chess champion, in a six-game match. To most people, beating Kasparov at chess would seem a far more impressive feat than coming up with "Famous First Names," say, or "State Birds." But chess is a game of strictly defined rules.

The open-endedness of "Jeopardy!"—indeed, its very goofiness—made it, for a machine, much more daunting.

Lickel's idea was batted around, rejected, and finally resurrected. In 2006, the task of building an automated "Jeopardy!" champion was assigned to a team working on question-answering technology, or QA. As Stephen Baker recounts in his book about the project, "Final Jeopardy," progress was, at first, slow. Consider the following (actual) "Jeopardy!" clue: "In 1984, his grandson succeeded his daughter to become his country's Prime Minister." A person can quickly grasp that the clue points to the patriarch of a political family and, with luck, summon up "Who is Nehru?" For a computer, the sentence is a quagmire. Is what's being sought a name? If so, is it the name of the grandson, the daughter, or the Prime Minister? Or is the question about geography or history?

Watson—basically a collection of processing cores—could be loaded with whole Wikipedias' worth of information. But just to begin to search this enormous database Watson had to run through dozens of complicated algorithms, which his programmers referred to as his "parsing and semantic analysis suite." This process yielded hundreds of "hypotheses" that could then be investigated.

After a year, many problems with Watson had been solved, but not the essential one. The computer took hours to generate answers that Jennings could find in an instant.

A year turned into two and then three. Watson's hardware was upgraded. Benefitting from algorithms that allowed him to learn from his own mistakes, he became more proficient at parsing questions and judging the quality of potential answers. In 2009, I.B.M. began to test the machine against former, sub-Jennings "Jeopardy!" contestants. Watson defeated some, lost to others, and occasionally embarrassed his creators. In one round, in response to a question about nineteenth-century British literature, the computer proposed the eighties pop duo Pet Shop Boys when the answer was Oliver Twist. In another round, under the category "Just Say No," Watson offered "What is fuck?" when the right response was "What is *nein*?"

I.B.M.'s aspirations for Watson went way beyond game shows. A computer that could cope with the messiness and the complexity of English could transform the tech world; one that could improve his own performance in the process could upend nearly everything else. Firms like Google, Microsoft, and Amazon were competing with I.B.M. to dominate the era of intelligent machines, and they continue to do so. For the companies involved, hundreds of billions of dollars are at stake, and the same could also be said for the rest of us. What business will want to hire a messy, complex carbon-based life form when a software tweak can get the job done just as well?

Ken Jennings, who might be described as the first person to be rendered redundant by Watson, couldn't resist a dig at his rival when the two finally, as it were, faced off. In January, 2011, Jennings and another former champion, Brad Rutter, played a two-game match against the computer, which was filmed in a single day. Heading into the final "Final Jeopardy!," the humans were so far behind that, for all intents and purposes, they were finished. All three contestants arrived at the correct response to the clue, which featured an obscure work of geography that inspired a nineteenth-century novelist. Beneath his answer—"Who is Bram Stoker?"—Jennings added a message: "I for one welcome our new computer overlords."

How long will it be before you, too, lose your job to a computer? This question is taken up by a number of recent books, with titles that read like variations on a theme: "The Industries of the Future," "The Future of the Professions," "Inventing the Future." Although the authors of these works are employed in disparate fields—law, finance, political theory—they arrive at more or less the same conclusion. How long? Not long.

"Could another person learn to do your job by studying a detailed record of everything you've done in the past?" Martin Ford, a software developer, asks early on in "Rise of the Robots: Technology and the Threat of a Jobless Future" (Basic Books). "Or could someone become proficient by repeating the tasks you've already completed, in the way that a student might take practice tests to prepare for an exam? If so, then there's a good chance that an algorithm may someday be able to learn to do much, or all, of your job."

Later, Ford notes, "A computer doesn't need to replicate the entire spectrum of your intellectual capability in order to displace you from your job; it only needs to do the specific things you are paid to do." He cites a 2013 study by researchers at Oxford, which concluded that nearly half of all occupations in the United States are "potentially automatable," perhaps within "a decade or two." ("Even the work of software engineers may soon largely be computerisable," the study observed.)

The "threat of a jobless future" is, of course, an old one, almost as old as technology. The first, rudimentary knitting machine, known as a "stocking frame," was invented in the late sixteenth century by a clergyman named William Lee. Seeking a patent for his invention, Lee demonstrated the machine for Elizabeth I. Concerned about throwing hand-knitters out of work, she refused to grant one. In the early nineteenth century, a more sophisticated version of the stocking frame became the focus of the Luddites' rage; in towns like Liversedge and Middleton, in northern England, textile mills were looted. Parliament responded by declaring "frame breaking" a capital offense, and the machines kept coming. Each new technology displaced a new cast of workers: first knitters, then farmers, then machinists. The world as we know it today is a product of these successive waves of displacement, and of the social and artistic movements they inspired: Romanticism, socialism, progressivism, Communism.

Meanwhile, the global economy kept growing, in large part *because* of the new machines. As one occupation vanished, another came into being. Employment migrated from farms and mills to factories and offices to cubicles and call centers.

Economic history suggests that this basic pattern will continue, and that the jobs eliminated by Watson and his ilk will be balanced by those created in enterprises yet to be imagined—but not without a good deal of suffering. If nearly half the occupations in the U.S. are "potentially automatable," and if this could play out within "a decade or two," then we are looking at economic disruption on an unparalleled scale. Picture the entire Industrial Revolution compressed into the life span of a beagle.

And that's assuming history repeats itself. What if it doesn't? What if the jobs of the future are also potentially automatable?

"This time is always different where technology is concerned," Ford observes. "That, after all, is the entire point of innovation."

Jerry Kaplan is a computer scientist and entrepreneur who teaches at Stanford. In "Humans Need Not Apply: A Guide to Wealth and Work in the Age of Artificial Intelligence" (Yale), he notes that most workplaces are set up to suit the way people think. In a warehouse staffed by people, like items are stored near one another—mops next to brooms next to dustpans—so their location is easy for stock clerks to remember. Computers don't need such mnemonics; they're programmed to know where things are. So a warehouse organized for a robotic workforce can be arranged according to entirely different principles, with mops, say, stored next to glue guns because the two happen to be often ordered together.

"When most people think about automation, they usually have in mind only the simple replacement of labor or improving workers' speed or productivity, not the more extensive disruption caused by process reengineering," Kaplan writes. Process reëngineering means that, no matter how much the warehouse business expands, it's not going to hire more humans, because they'll just get in the way. It's worth noting that in 2012 Amazon acquired a robotics company, called Kiva, for three-quarters of a billion dollars. The company's squat orange bots look like microwave ovens with a grudge. They zip around on the ground, retrieving whole racks' worth of merchandise. Amazon now deploys at least thirty thousand of them in its fulfillment centers. Speaking of the next wave of automation, Amazon's chairman, Jeff Bezos, said recently, "It's probably hard to overstate how big of an impact it's going to have on society over the next twenty years."

Not long ago, a team of researchers at Berkeley set out to design a robot that could fold towels. The machine they came up with looked a lot like Rosie, the robot maid on "The Jetsons," minus the starched white cap. It had two cameras mounted on its "head" and two more between its arms. Each arm could rotate up and down and also sideways, and was equipped with a pincer-like "gripper" that could similarly rotate. The robot was supposed to turn a mess of towels into a neat stack. It quickly learned how to grasp the towels but had a much harder time locating the corners. When the researchers tested the robot on a pile of assorted towels, the results were, from a practical standpoint, disastrous. It took the robot an average of twenty-four and a half minutes to fold each towel, or ten hours to produce a stack of twenty-five.

Even as robots grow cleverer, some tasks continue to bewilder them. "At present, machines are not very good at walking up stairs, picking up a paper clip from the floor, or reading the emotional cues of a frustrated customer" is how the M.I.T. researchers Erik Brynjolfsson and Andrew McAfee put it, in "The Second Machine Age: Work, Progress, and Prosperity in a Time of Brilliant Technologies" (Norton). Because we see the world through human eyes and feel it with human hands, robotic frustrations are hard for us to understand. But doing so is worth the effort, Brynjolfsson and McAfee contend, because machines and their foibles explain a lot about our current economic situation.

Imagine a matrix with two axes, manual versus cognitive and routine versus nonroutine. Jobs can then be arranged into four boxes: manual routine, manual nonroutine, and so on. (Two of Brynjolfsson and McAfee's colleagues at M.I.T., Daron Acemoglu and David Autor, performed a formal version of this analysis in 2010.) Jobs on an assembly line fall into the manual-routine box, jobs in home health care into the manual-nonroutine box. Keeping track of inventory is in the cognitive-routine box; dreaming up an ad campaign is cognitive nonroutine.

The highest-paid jobs are clustered in the last box; managing a hedge fund, litigating a bankruptcy, and producing a TV show are all cognitive and nonroutine. Manual, nonroutine jobs, meanwhile, tend to be among the lowest paid—emptying bedpans, bussing tables, cleaning hotel rooms (and folding towels). Routine jobs on the factory floor or in payroll or accounting departments tend to fall in between. And it's these middle-class jobs that robots have the easiest time laying their grippers on.

During the recent Presidential campaign, much was said—most of it critical—about trade deals like the North American Free Trade Agreement and the Trans-Pacific Partnership. The argument, made by both Bernie Sanders and Donald Trump, was that these deals have shafted middle-class workers by encouraging companies to move jobs to countries like China and Mexico, where wages are lower. Trump has vowed to renegotiate NAFTA and to withdraw from the T.P.P., and has threatened to slap tariffs on goods manufactured by American companies overseas. "Under a Trump Presidency, the American worker will finally have a President who will protect them and fight for them," he has declared.

According to Brynjolfsson and McAfee, such talk misses the point: trying to save jobs by tearing up trade deals is like applying leeches to a head wound. Industries in China are being automated just as fast as, if not faster than, those in the U.S. Foxconn, the world's largest contract-electronics company, which has become famous for its city-size factories and grim working conditions, plans to automate a third of its positions out of existence by 2020. The *South China Morning Post* recently reported that, thanks to a significant investment in robots, the company already has succeeded in reducing the workforce at its plant in Kunshan, near Shanghai, from a hundred and ten thousand people to fifty thousand. "More companies are likely to follow suit," a Kunshan official told the newspaper.

"If you look at the types of tasks that have been offshored in the past twenty years, you see that they tend to be relatively routine,"

"During his 'victory lap' through the Midwest, the President-elect vowed to 'usher in a new Industrial Revolution,' apparently unaware that such a revolution is already under way, and that this is precisely the problem."

Brynjolfsson and McAfee write. "These are precisely the tasks that are easiest to automate." Off-shoring jobs, they argue, is often just a "way station" on the road to eliminating them entirely.

In "Rise of the Robots," Ford takes this argument one step further. He notes that a "significant 'reshoring' trend" is now under way. Reshoring reduces transportation costs and cuts down on the time required to bring new designs to market. But it doesn't do much for employment, because the operations that are moving back to the U.S. are largely automated. This is the major reason that there is a reshoring trend; salaries are no longer an issue once you get rid of the salaried. Ford cites the example of a factory in Gaffney, South Carolina, that produces 2.5 million pounds of cotton yarn a week with fewer than a hundred and fifty workers. A story about the Gaffney factory in the *Times* ran under the headline "U.S. TEXTILE PLANTS RETURN, WITH FLOORS LARGELY EMPTY OF PEOPLE."

As recently as twenty years ago, Google didn't exist, and as recently as thirty years ago it *couldn't* have existed, since the Web didn't exist. At the close of the third quarter of 2016, Google was valued at almost five hundred and fifty billion dollars and ranked as the world's second-largest publicly traded company, by market capitalization. (The first was Apple.)

Google offers a vivid illustration of how new technologies create new opportunities. Two computer-science students at Stanford go looking for a research project, and the result, within two decades, is worth more than the G.D.P. of a country like Norway or Austria. But Google also illustrates how, in the age of automation, new wealth can be created without creating new jobs. Google employs about sixty thousand workers. General Motors, which has a tenth of the market capitalization, employs two hundred and fifteen thousand people. And this is G.M. post-Watson. In the late nineteen-seventies, the carmaker's workforce numbered more than eight hundred thousand.

How much technology has contributed to the widening income gap in the U.S. is a matter of debate; some economists treat it as just one factor, others treat it as the determining factor. In either case, the trend line is ominous. Facebook is worth two hundred and seventy billion dollars and employs just thirteen thousand people. In 2014, Facebook acquired Whatsapp for twenty-two billion dollars. At that point, the messaging firm had a grand total of fifty-five employees. When a twenty-two-billion-dollar company can fit its entire workforce into a Greyhound bus, the concept of surplus labor would seem to have run its course.

Ford worries that we are headed toward an era of "techno-feudalism." He imagines a plutocracy shut away "in gated communities or in elite cities, perhaps guarded by autonomous military robots and drones." Under the old feudalism, the peasants were exploited; under the new arrangement, they'll merely be superfluous. The best we can hope for, he suggests, is a collective form of semi-retirement. He recommends a guaranteed basic income for all, to be paid for with new taxes, levelled, at least in part, on the new gazillionaires.

To one degree or another, just about everyone writing on the topic shares this view. Jerry Kaplan proposes that the federal government create a 401(k)-like account for every ten-year-old in the U.S. Those who ultimately do find jobs could contribute some of their earnings to the accounts; those who don't could perform volunteer work in return for government contributions. (What the volunteers would live off is a little unclear; Kaplan implies that they might be able to get by on their dividends.) Brynjolfsson and McAfee prefer the idea of a negative income tax; this would provide the unemployed with a minimal living and the underemployed with additional cash.

But, if it's unrealistic to suppose that smart machines can be stopped, it's probably just as unrealistic to imagine that smart policies will follow. Which brings us back to Trump. The other day, during his "victory lap" through the Midwest, the President-elect vowed to "usher in a new Industrial Revolution," apparently unaware that such a revolution is already under way, and that this is precisely the problem. The pain of dislocation he spoke to during the campaign is genuine; the solutions he offers are not. How this will all end, no one can say with confidence, except, perhaps, for Watson.

LET THEM DRINK BLOOD

A. M. Gittlitz

(First appeared in The New Inquiry, *December 27, 2016, "Science and/or Fiction" Issue)*

Silicon Valley's elites are a revolutionary vanguard party developing the not-too-distant future of cybernetic capitalist reconstruction. Despite cultish personas and massive social influence, however, they tend to keep their politics on the low. That changed this year when Peter Thiel, PayPal founder and Facebook board member, who also has investments in SpaceX and data analysis firm Palantir, revealed himself as mastermind of the litigious assassination of Gawker, a fellow-traveler of right-libertarian White Nationalists, and a prominent supporter of President-elect Donald J. Trump.

Thiel's "Don't Be Evil" competitors now look like saints in comparison. Some colleagues distanced themselves, while others wrote off the endorsement as part of his "disruptive instinct" to break down regulations preventing his Founders Fund investments from expanding. Then, in August, it was rumored that Thiel bragged to friends that Trump promised to nominate him to the Supreme Court, which would make him one of the most powerful men in America for a lifetime term. And Peter Thiel plans to live for a long time. He has a personal and financial stake in life extension technologies, including "parabiosis"—the (theoretically) rejuvenating transfer of young blood to an older person.

For those outside the valley, Thiel's vampiric ambitions appeared to vindicate populist imagery dating back to Voltaire, who wrote in his *Philosophical Dictionary* that the real vampires were "stock-jobbers, brokers, and men of business, who sucked the blood of the people in broad daylight." A century of trite political cartoons have depicted moguls or aristocrats growing fat on the blood of innocents. Most recently, Matt Taibbi's popular description of Goldman Sachs as a "vampire squid wrapped around the face of humanity," revivified this discourse as a conceptual rallying point of Occupy Wall Street. It was a sentiment that even played out in the campaigns of Sanders, and to a far worse extent, Trump, who towards the end of his campaign regularly parroted Infowars radio host Alex Jones's discourse about a world-dominating conspiracy of shadowy globalists.

"Elites around the world have been obsessed with blood for thousands of years," Jones said in an Infowars video this summer concerning Thiel. He goes on to argue that elites throughout history, including the British Royal Family, have undergone similar parabiotic treatments for decades. "Where the story really gets weird," Jones opined, "is that Prince Charles came out in the last decade and said I am a direct descendent of Vlad the Impaler... the people running things aren't physical, immortal vampires, but they have the spirit of what you describe as a vampire, and they believe their god, Lucifer, if they establish a world government, is going to give them eternal life. And now they're mainlining the idea of baby parts and blood from the young to make the rich live longer."

Dropping in Dracula's relation to the British Monarchy would be irrelevant for a journalist, but for a conspiracist like Jones, the detail is delicious enough to aid both his legitimate thesis—that the rich and powerful treat the world's populations as nothing but commodities—and the farfetched one: Thiel, despite being a fellow traveler of Jones' paleoconservatism, is an early adopter of technology that would free him from the eternal hellfire he would otherwise be due through his deals with the devil. Jones warns that Thiel's fellow globalists will continue to push wars, cancer-causing vaccines, and abortions in a eugenic blood-ritual to depopulate the world by 80% and install a one-world government.

Thiel's visionary investments suggest a similar blurring of science fiction, paranoia, and plausible dystopian scenarios. In a 2009 essay for *Cato Unbound*, he stated his anti-national principles: "I stand against confiscatory taxes, totalitarian collectives, and the ideology of the inevitability of the death of every individual."

So what's standing in the way of a "death and taxes"-optional world? The same thing that fellow frontier industrialist Daniel Plainview lamented in 2007's *There Will Be Blood*: People. Poor and female ones, specifically. "Since 1920," Thiel continued, "the vast increase in welfare beneficiaries and the extension of the franchise to women—two constituencies that are notoriously tough for libertarians—have rendered the notion of 'capitalist democracy' into an oxymoron."

"Because there are no truly free places left in our world, I suspect that the mode for escape must involve some sort of new and hitherto untried process that leads us to some undiscovered country; and for this reason I have focused my efforts on new technologies that may create a new space for freedom." He outlines three such spaces: artificial island micronations, the Internet, and space colonies. He's invested heavily in all three, including SpaceX, which plans to colonize Mars. Now, he's a member of Trump's transition team. In a potential sign of Thiel's influence in the new administration, the President-elect's senior adviser on NASA recently announced that funds for studying climate change will be diverted to deep space exploration.

Thiel's plan to locate and conquer space to build a libertarian utopia closely follows the *deus ex machina* of *Atlas Shrugged*—a perpetual motion machine that allowed the libertopia Galt's Gulch to become a fully-automated bourgeois paradise without need for whiney workers. But even if the pretense of these microstates were egalitarian (and not the aristocratic *Reichlets* described by Thiel's mentor and lead thinker of the *Property and Freedom Society*, right-libertarian economist Hans-Hermann Hoppe), the *Communist Manifesto*'s critique of utopian ambition in literature is fitting:

> They still dream of experimental realization of their social Utopias, of founding isolated "phalansteries," of establishing "Home Colonies," or setting up a "Little Icaria," duodecimo editions of the New Jerusalem. By degrees, they sink into the category of the reactionary [or] conservative Socialists depicted above, differing from these only by more systematic pedantry, and by their fanatical and superstitious belief in the miraculous effects of their social science.
>
> They, therefore, violently oppose all political action on part of the working class; such action, according to them, can only result from blind unbelief in the new Gospel.

Before Thiel and the broader space-race-boosterism science fiction canon, futurology was a misguided socialist enterprise as described by Marx. Some of Thiel's aspirations were even disastrously attempted in the 20s and 30s by two Bolshevik factions calling themselves the God-Builders and Biocosmist-Immortalists. The writers Gorky, Lunacharsky, Malevich, and Bogdanov were amongst their ranks, and all inspired by Russian Orthodox philosopher and mystic Nikolai Fyodorov, who advocated life extension and space colonization in technological culmination of the Book of Revelation. When Lenin died, Vladimir Mayakovsky's declaration of "Lenin lived, Lenin lives, Lenin will live forever!" represented the sentiments of his peers, who preserved his body, organs, and brain, in hopes they could revive him.

Alexander Bogdanov, a cofounder of the Bolshevik party and Lenin's one-time rival, was a particularly Thielian figure in the group. In 1905, Bogdanov wrote the science fiction novel *Red Star*, depicting a communist society on Mars where parabiosis was practiced as a form of mutual aid. Two decades after writing *Red Star*, Bogdanov founded the State Institute for Haematology and Blood Transfusions. He subjected himself to these transfusions, and died from a botched trial—a fitting metaphor for a revolutionary killed by his own "fanatical and superstitious belief in the miraculous power of their social sciences."

By the 1930's, some remaining Immortalists were making the case for Stalinist terror. Writing for *Pravda*, Gorky described peasants resisting collectivization and their starving orphaned children as "masses of parasites…rats, mice, gophers," who must be wiped out because they "do the economy of the country a great deal of harm." In his book *The Immortalization Commission*, contemporary philosopher John Gray argues that extermination fit well with the transhumanist foundation of God-Building. Following Lenin's promise that the peasants will "worship electricity," God-Builders believed once industrialized, the Soviet Union would advance its technological capabilities to fulfill the Christian eschatology by liberating the New Soviet man from the constraints of mortality, terrestriality, and embodiment. Similar to Kurzweil's "singularity," Gray calls it a "materialist rapture" in which:

> The dead will be resurrected by the power of science. Severing their links with the flesh, humans will enter a deathless realm. Lower life forms—plants, animals and unregenerate humans—will be left behind, or else eradicated. All that will remain will be the "pure thought" Gorky envisioned in his conversation with Blok—infinite, immortal energy.

In a way, Gorky was right. During the years of the New Economic Policy (1921–1928), state power centralized to a point that its revolutionary goals only existed in the "pure thought" of elevated apparatchiks. Outside the party, political obstacles were easily dealt with by the secret police, and the theory of Socialism in One

"For the middle class, extra years of life will be purchasable in mortgage-like installments... Dying will increasingly be viewed as a manageable epidemic, like AIDS, violent crime, or homelessness."

Country turned the Soviet Union into a *SimCity* terrain to be built and destroyed as Stalin pleased. Although the Bolsheviks formed their party in a reaction to the horrors of industrial capital and World War I, it's no surprise that the cruelty and terror used to consolidate their power resulted in a totalitarian society.

Thiel views the world much like the early Soviet futurists. Their utopian dreams ran far ahead of the chaos of revolutionary Russia, where Civil War and social upheaval posed a significant impediment to the development of the Soviet Union's productive forces. Our own era of bicameral stagnation, social unrest, and organized labor similarly threaten the reactionary acceleration envisioned by Silicon Valley futurists, who are developing technology to eliminate rebellion through expansion of the carceral state, scientific breakthroughs to protect the wealthiest from irreversible environmental depletion, and a new relationship between life and death mediated by the dead labor of capitalism.

Organs, blood, or stem cells may soon be freely traded like an *Uber* for *Sein-zum-Tode* (although, with scant evidence that life extension is anything other than pseudoscience, it's more likely to be a *Theranos* for *Thanatos*). For the middle class, extra years of life will be purchasable in mortgage-like installments. Life extension will be distributed just like the resiliency plans of major population areas under the menace of natural disasters amplified by global warming. The wealthiest areas will fortify structures, raise sea-walls, and afford for evacuations, while places like Haiti and Bangladesh are doomed to drown. Dying will increasingly be viewed as a manageable epidemic, like AIDS, violent crime, or homelessness.

Mars is even more open to the whims of venture capitalists who talk about it as if it's a cold red stress-ball for the worst mistakes of humanity. The most commonly discussed technique for making the planet habitable involves exporting global warming to Mars by building robotic factories that produce nothing but greenhouse gases, thus melting the ice caps and making the atmosphere more like that of Earth. Elon Musk had one other idea: nuking it.

If all of Silicon Valley were revealed to be drinking plasma instead of pinot, so what? Historically, it has not only been the elite that drink blood. Medieval historian Richard Sugg recently told *Smithsonian Magazine* that villagers would gather around the recently executed with bowls to drink their blood fresh, or congeal it into a pudding for later. "The executioner was seen as a big healer," Sugg said. The 2016 election was such a ritual, meant to unite the people and reify the power of the Sovereign. A Clintonian decapitation of Trumpism would have reassured America that "Trump is not who we are," even as Obama's immigrant detention camps remain full past capacity and Kissingerian quagmires continue to burn across the globe. Instead, it is Clinton, utterly exposed in all her hypocrisy, who will face the new regime's pillory alongside the entirety of the Muslim and undocumented population.

But it wasn't Trump's "they all must go" eliminationism that lured Thiel, nor the unlikely Supreme Court nod. According to colleagues interviewed by *Bloomberg*, it was his Silicon Valley disruptor instincts—speculating that Trump will return the favor for Thiel's endorsement by helping Palantir with a government contract and giving SpaceX a leg-up against Boeing. With control over big data, the economics of life and death, and his own sandbox planet to build as he sees fit, Thiel is positioning himself beyond critique or recall from the masses.

Nonetheless, like the rest of his class, Thiel will always serve a higher power. For Marx, it is not the capitalists who are vampires, but "the thing they represent," the non-human force, a "dead labor" which "lives only by sucking living labor, and lives the more, the more labor it sucks." Painting Thiel as a vampire may be comforting, because then it would only take some sunlight and a pointy piece of wood to take him down. Instead, consider what Thiel really fears: *us*, and our historical tendency to commit deicide.

BLACK AMERICANS AND ENCRYPTION: THE STAKES ARE HIGHER THAN APPLE V. FBI

Malkia Cyril

(First appeared in The Guardian, *March 21, 2016)*

When the FBI branded Martin Luther King Jr a "dangerous" threat to national security and began tapping his phones, it was part of a long history of spying on black activists in the United States. But the government surveillance of black bodies has never been limited to activists—in fact, according to the FBI, you only had to be black.

In the 2016 legal battle between Apple and the FBI, black perspectives were largely rendered invisible, yet these communities stood to lose big. A federal judge was asked by the FBI to compel Apple to unlock the iPhone of a San Bernardino shooter whose phone data had been encrypted. After finding a third party to crack the phone, the DOJ pulled their request and the judge was never forced to rule on the case.

While seemingly about protecting national security—the same rationale used to justify 20th century surveillance of MLK, the Black Panther Party and others—this case is about much more. It could establish a legal precedent used to suppress the growing movement for black lives that is deposing public officials and disrupting the daily assault on black people in cities across the country.

Building off the civil rights and black power movements of the 1960s, a 21st century movement for black lives is coming of age, mobilizing the same courageous methods of non-violent direct action, using the same local-to-local strategy, and making many of the same demands. An intersectional approach is replacing old identity politics, and a newfound digital landscape is making communication possible in more directions and at previously unimaginable speeds. The movement for black lives is attracting the brightest minds and bravest bodies. Black activists are developing new ways of grassroots organizing in an information economy.

Like its predecessors, the movement for black lives has been met by anti-democratic state surveillance and anti-black police violence. New "smart" policing methods are being used by modern-day gumshoes who, fueled by the false rhetoric of black criminality, experiment with high-tech tools to the detriment of black democratic engagement.

In the 20th century, the FBI admitted to overreaching and violating the constitution when it used its counter intelligence program, COINTELPRO, for domestic surveillance that spied on black activists. Last year, FBI director James Comey admitted in a congressional committee hearing to flying spy planes that monitored protests in the wake of police killings of black people in Ferguson and Baltimore with the agency working in tandem with local police. In Chicago, home of the infamous "red squad," police collected "First Amendment Worksheets" on black organizations such as We Charge Genocide, and Jesse Jackson's Rainbow Push Coalition.

There are reports from activists on the front lines of protests about police employing "kill switch" technology to cut off live-streaming, using Stingrays to intercept phone calls, or flying drones overhead for crowd control, but such claims are unconfirmed as police refuse to reveal their techniques and are not compelled by law to do so.

Twentieth century surveillance is alive and well in the 21st century, and is one powerful reason why, in a digital age and era of big data, the fight for racial justice must also include a fight for the equal and fair application of first and fourth amendment rights.

A letter was sent by some of us in the Black Lives Matter movement to California federal magistrate judge Sheri Pym, who oversaw the Apple case, warning of the dangerous implications of siding with the FBI. It was signed by Beats, Rhymes & Relief, the Gathering for Justice, Justice League NYC, writer Shaun King, Black Lives Matter co-founder and Black Alliance for Just Immigration executive director Opal Tometi, as well as the organization I work for, the Center for Media Justice.

I signed because, as the child of a Black Panther, I grew up with the persistent threat of police spying. The police "watched" my family in the name of "safety" and "national security," but I knew that we became targets of government surveillance because my mother advocated for black bodies that had been abandoned and abused by state violence.

That is why the FBI case was not only against Apple, but was also against communities of color and communities of resistance. It was against democracy. It was against the black immigrant worker who has fled political persecution, the black and Latino youth putting themselves on the line to catalyze deep change, the gender non-conforming bodies subjected to daily assaults, the Muslim communities regularly targeted by bias and hate crimes. We don't have the same protections others take for granted, we are instead treated as perpetually guilty.

Reports have surfaced that the Department of Homeland Security has been monitoring the movement for black lives since the initial uprisings in Ferguson. We know that police are watching the tweets we write, the Facebook event pages we set up, and the demonstrations we organize in the streets. If we are arrested, our phones will be confiscated. Whether or not police can look into our phones, whether or not they need a warrant, is being tested in court. This is not a vision of some distant dystopic future, this is happening right now. This is why the FBI case against Apple was also against us.

"Democratic engagement and the exercise of our human and civil rights in a digital age demands the ability to encrypt our communications."

For black communities and others pushed to the margins of political and economic power—democratic engagement and the exercise of our human and civil rights in a digital age demands the ability to encrypt our communications.

It isn't just black activists either—Latino activists are raising a similar rallying cry. Consider the possibilities under President Trump, who has notoriously expressed his anti-immigrant views, and sided with the FBI in its fight against Apple. With record numbers of deportations already at hand—could undocumented immigrants be rounded up using the information transmitted from their cellphones?

A newly-developed open source app for iPhones called Signal, which encrypts phone calls and text messages, has become a favorite among organizers, as well as Edward Snowden. It allows for free and instant encryption of messages that cannot be cracked by anybody wanting to eavesdrop. Activists across the world have adopted the app as one way to protect their right to organize.

Yet encryption technology is for more than just activists. Whether protecting from identity theft or government surveillance—all communities need to protect their data in the digital age. We cannot have a healthy democracy without everyone's voice.

Black voices, and other voices of color, have long been missing from the debates on government surveillance—but not anymore. We're here, and we are calling on companies to protect the rights of consumers, and on legislators to protect the rights of residents. One way we did so was to support the Encrypt Act of 2016, legislation in the House of Representatives, which would have prevented the government, or a contracted company, from altering the security functions of computers and cellphones, or decoding encrypted information, in order to conduct a search.

Campaigns to protect encryption have been answered by legislation intended to put the squeeze on smartphone makers. A few months later, a draft bill called the Compliance with Court Orders Act of 2016 was released in a bipartisan effort by Senators Dianne Feinstein (D-CA) and Richard Burr (R-NC). The language, while not explicitly outlawing encryption, required tech companies to render data "intelligible" if compelled by a judge's order.

We are also seeing legislation designed to ban encryption at the state level. In 2015, an anti-encryption bill was introduced in New York, and the following year a similar bill was floated in California, both using the rhetoric of terrorism and black criminality. If bills like these were passed, they would force companies to manufacture smartphones compliant with laws in two of the largest markets in the nation.

Soon after Trump came into office, there was a repeal of 2016 regulations limiting what broadband providers could do with customer data. Since then, at least 20 states have advanced legislation to restore privacy rights. In California, AB 375, introduced by Assemblyman Ed Chau (D–Monterey Park), would require Internet service providers like Verizon, Comcast, and AT&T to get permission from a customer before handing over his or her data.

Encryption is also a global debate. In June 2017, officials from the "Five Eyes"—a surveillance partnership of intelligence agencies from Australia, Canada, New Zealand, the United Kingdom, and the United States—met in Ottawa to discuss digital security issues, including encryption. In response, a five-nation coalition of NGOs and security experts sent a letter to the Five Eyes governments asking them not to mandate "back doors" in the technology and systems that we rely on to keep us safe. They recommended a "multi-stakeholder forum" that would involve public participation and ensure the protection of human rights.

Communities of color and poor people are especially vulnerable. Not all smartphones are created equal. In the US and across the world, wealthier consumers tend to buy more expensive iPhones, which have better security protections. Android phones can be purchased for as little as $50, and are more popular among working-class people and African Americans, but can be easily hacked. The digital divide impacts who has the most privacy. Indeed, this issue is not just about privacy, but about social and political autonomy, which only exists for those with enough power or privilege to preserve it.

In this fight, tech companies are the key. If they align with authoritarianism, we have everything to fear. But if they align with human rights, democracy may have found the giant it needs to push back against these state attacks on encryption. The Center for Media Justice and those we align with are working to bring tech companies to the table to remind them of the deeper ramifications of their decisions for communities of color and oppressed peoples.

Encryption is necessary for black civil and human rights to prosper, but isn't enough. While it protects our democratic right to organize for change, we must fight for a world in which those rights are not under persistent threat. The Apple v. FBI case was a test case for democracy. Who owns the media will determine, for this and the next generation, who has the right to communicate, and therefore the power to define reality.

In the encryption debate, the stakes are high for black people. Indeed, we are in a fight for our lives. I believe that we will win.

POLICING THE FUTURE: IN THE AFTERMATH OF FERGUSON, ST. LOUIS COPS EMBRACE CRIME-PREDICTING SOFTWARE

Maurice Chammah
(with additional reporting by Mark Hansen)

(First appeared in The Verge*; story produced in partnership with the Marshall Project, February 3, 2016)*

Just over a year after Michael Brown's death became a focal point for a national debate about policing and race, Ferguson and nearby St. Louis suburbs had returned to what looked, from the outside, like a kind of normalcy. Near the Canfield Green apartments, where Brown was shot by police officer Darren Wilson, a sign reading "Hands Up Don't Shoot" and a mountain of teddy bears had been cleared away. The McDonald's on West Florissant Avenue, where protesters nursed rubber bullet wounds and escaped tear gas, was now just another McDonald's.

Half a mile down the road in the city of Jennings, between the China King restaurant and a Cricket cell phone outlet, sat an empty room that the St. Louis County Police Department keeps as a substation. During the protests, it had been a war room, where law enforcement leaders planned their responses to the chaos outside.

One day in December 2015, a few Jennings police officers flicked on the substation's fluorescent lights and gathered around a big table to eat sandwiches. The conversation drifted between the afternoon shift's mundane roll of stops, searches, and arrests, and the day's main excitement: the officers were trying out a new software program called HunchLab, which crunches vast amounts of data to help predict where crime will happen next.

The conversation also turned to the grand anxieties of post-Ferguson policing. "Nobody wants to be the next Darren Wilson," Officer Trevor Voss told me. They didn't personally know Wilson. Police jurisdiction in St. Louis is notoriously labyrinthine and includes dozens of small, local municipal agencies like the Ferguson Police Department, where Wilson worked—munis, the officers call them—and the St. Louis County Police Department, which patrols areas not covered by the munis and helps with "resource intense events," like the protests in Ferguson. The munis have been the targets of severe criticism; in the aftermath of 2014's protests, Ferguson police were accused by the federal Department of Justice of being racially discriminatory and poorly trained, more concerned with handing out tickets to fund municipal coffers than with public safety.

The officers in Jennings work for the St. Louis County Police Department; in 2014, their colleagues appeared on national TV, pointing sniper rifles at protesters from armored trucks. Since then, the agency has also been called out by the Justice Department for, among other things, its lack of engagement with the community.

Still, the county police enjoy a better local reputation than the munis. Over the last five years, Jennings precinct commander Jeff Fuesting has tried to improve relations between officers—nearly all white—and residents—nearly all black—by going door to door for "Walk and Talks." Fuesting had expressed interest in predictive policing years before, so when the department heads brought in HunchLab, they asked his precinct to roll it out first. They believed that data could help their officers police better and more objectively. By identifying and aggressively patrolling "hot spots," as determined by the software, the police wanted to deter crime before it ever happened.

HunchLab, produced by Philadelphia-based startup Azavea, represents the newest iteration of predictive policing, a method of analyzing crime data and identifying patterns that may repeat into the future. HunchLab primarily surveys past crimes, but also digs into dozens of other factors like population density; census data; the locations of bars, churches, schools, and transportation hubs; schedules for home games—even moon phases. Some of the correlations it uncovers are obvious, like less crime on cold days. Others are more mysterious: rates of aggravated assault in Chicago have decreased on windier days, while cars in Philadelphia were stolen more often when parked near schools.

At the same time, a growing chorus of activists and academics worry that the reliance on data is a sign that police departments have not adequately heeded the lessons of Ferguson. Kade Crockford, the director of the Technology for Liberty program at the Massachusetts ACLU, says that predictive policing is based on "data from a society that has not reckoned with its past," adding "a veneer of technological authority" to policing practices that still disproportionately target young black men. In other cities, some police departments are even moving toward predicting which people, rather than which places, are most crime-prone.

"At a time when communities are crying out for justice," Crockford told me, "I never heard anyone in one of these communities say, 'I think police need to use more computers!'"

Predicting crime has always been part of police work; any beat cop can tell you that a particularly dark street corner is vulnerable to carjackers, or a large parking lot offers anonymity for drug dealers. Scholars have been mapping crime since the 1800s, but during New York City's crime spike in the 1990s, police officers started doing so systematically. Most notable among them was Jack Maple, a quick-talking, up-from-the-bottom transit cop who wore double-breasted suits, homburg hats, and two-tone shoes and has become a near-mythic figure in police circles. At the NYPD's Manhattan headquarters, Maple would stretch out butcher paper across 55 feet of wall space. "I called them the Charts of the Future," he once told an interviewer. "I mapped every train station in New York City and every train. Then I used crayons to mark every violent crime, robbery, and grand larceny that occurred."

Maple's boss, Police Commissioner Bill Bratton, sent officers to patrol the areas Maple marked up. The process evolved into an entire system of police management called CompStat, which uses data to hold individual precinct commanders accountable for the crime levels in their areas. In varying forms, "hot-spot policing" has spread throughout the nation's police departments. Bratton calls it "computerized fishing."

"Cops-on-dots," as it's sometimes known, has often been associated with Bratton's other major legacy, "Broken Windows," in which police target low-level offenses like graffiti and public drinking, creating a sense of public order that is believed to deter more serious crimes. Such tactics have been credited with helping bring down crime rates, but they have also contributed to the aggressive targeting—and stopping and searching—of black people, fostering resentment of police in many communities.

St. Louis officials had been using data to send police to patrol hot spots since 2009; today the city holds weekly meetings for commanders to discuss why certain crimes keep hitting certain places, and how to address it. When one precinct captain noticed a lot of robberies of appliances from houses under construction, officers were instructed to keep track of building schedules. In agencies across the country, the more commanders looked at the data, the more timely their responses to that data could be, and crime analysis started edging toward real time. The dream was to go beyond the present.

Throughout the criminal justice system, a faith in data's ability to improve upon human judgment has led judges, prosecutors, and other officials in recent years to embrace tools that address the future; many use "risk assessments" of defendants—which involve questionnaires about demographics, family, and personal history—in sentencing decisions. The White House has asked

Silicon Valley companies if they can develop algorithms to predict which people are likely to become "radicalized."

In the summer of 2014, a couple of months before Ferguson erupted, St. Louis County Police Chief Jon Belmar returned from a conference of police leaders in Boston, where he had been impressed by presentations from mathematicians and data analysts. He told his aide, Sgt. Colby Dolly, that he wanted their department to join dozens of cities already using predictive policing software.

As Dolly studied the predictive policing market, he found it was crowded with competitors. Since 2009, the National Institute of Justice had been funding research into crime prediction, transforming the field into big business. IBM, Hitachi, and Lexis had all begun to offer ways to predict crime through data.

The leader in the field is PredPol, a company that grew out of a team of researchers and officers working under Bratton during the chief's mid-2000s stint in Los Angeles. PredPol's algorithms digest years of data on crime locations, times, and types, spitting out the spots most likely to be hit by crime again. After using PredPol for four months, police in the Foothill Division in the San Fernando Valley claimed that property crime dropped 13 percent, while in the rest of the city, it rose by 0.4 percent. PredPol has received millions in venture capital funding and is now used by more than 50 police agencies in the US and UK.

But Dolly was attracted by Azavea's ability to analyze the impact of businesses, churches, and weather patterns on criminal activity. It was also cheaper: Azavea quoted around $50,000 for a year of HunchLab, where PredPol was asking for roughly $200,000.

Azavea's employees have a Silicon Valley ebullience—their website mentions "ping pong tournaments, team runs, hackathons," and "chess matches over lunch"—but they do not share the tech industry's talk of "disruption." Their rhetoric is civic-minded; the company's other projects include tools to analyze legislative districts, as well as an app that helps city residents map the locations of trees in order to study their environmental impact.

As predictive policing has spread, researchers and police officers have begun exploring how it might contribute to a version of policing that downplays patrolling—as well as stopping, questioning, and frisking—and focuses more on root causes of particular crimes. Rutgers University researchers specializing in "risk terrain modeling" have been using analysis similar to HunchLab to work with police on "intervention strategies." In one Northeast city, they have enlisted city officials to board up vacant properties linked to higher rates of violent crime, and to advertise after-school programming to kids who tend to gather near bodegas in high-risk areas.

Dolly was not opposed to examining and addressing the causes of crime, but the department was still focused on patrolling. He hoped using HunchLab might improve relations with the community by reducing the frequency with which police had to aggressively sweep an area in the wake of a crime. "You can only go so far in enforcing or arresting your way out of crime issues," he said. "This is a way to combat crime that should have minimal impact on the community."

In order to sidestep concerns about racially disproportionate policing, Dolly asked HunchLab to only predict the kinds of serious felonies that result in 911 calls, and not low-level crimes like drug possession. He asked the analysts to produce two boxes for every patrol area—no matter how wealthy or poor, black or white—showing two areas at the highest risk of crime for every 8-hour shift. But Dolly also recognized the fundamental limitation of the tool—it was "telling you where to go," he said. "It's not telling you what to do."

A few hours before police officers in Jennings started their afternoon patrol, Dolly sat down at his computer at police headquarters. He logged into the HunchLab website and pulled up a map. The sprawling metropolis was covered in little bright dots. He clicked to zoom in, and the dots grew into transparent boxes, each covering a space roughly half the size of a city block, and each tinted green, orange, red, purple, blue, pink, or yellow. The colors indicated which type of crime was most likely to hit that box: green for larceny, orange for gun crimes, red for aggravated assault.

As Dolly zoomed in on Jennings, he saw two boxes tinted green to indicate a high risk of larceny. He knew this area was one of Jennings' only commercial districts, so of course there would be a lot of shoplifting. As he panned toward the residential neighborhoods nearby, however, he saw red and orange boxes in areas that looked fairly random. "I've been doing police work 16 years," he said, "and I don't think you'd be able to isolate locations like this."

A few hours later, Thomas Keener arrived for his afternoon shift, checked his gun, got into his squad car, and pulled up the same map. Ten hours a day, four days a week, Keener's primary job is to answer 911 calls and provide backup to other officers. When calls don't come in, he patrols. Keener, 27, grew up in southern Missouri and graduated from the police academy six years ago. He is unfailingly polite. While his peers wear short sleeves, he chooses a long-sleeve khaki uniform and a dark brown tie, a formal get-up that, coupled with his buzzed hair, accentuates his boyishness.

When Keener began his shift, he headed toward the boxes HunchLab deemed to be high-risk. Like Dolly, he immediately registered that a green larceny box was over an area that contained a couple of dollar stores where he has caught people running out with stolen goods. He pointed out common escape routes. "See how it's easy to disappear over there?" Even without HunchLab, he would have probably gone to the area. In other cities where HunchLab has been used, police officers are often unsurprised by the locations of the boxes—police in Lincoln, Nebraska, started experimenting with the software in 2014 but have found it mostly tells them what they already know. "When I look at the HunchLab maps," said former police chief Tom Casady, "I say, 'Yep, it got that right!'"

The shift rolled on, and Keener got a series of calls: to help a man who had overdosed, to assist with an arrest at the police station, to help look for some young men in hoodies suspected of a burglary.

The day was proving to be a safe one. "I'm going to choose to credit the patrols with that," Keener said.

Driving through Jennings, it was clear Keener already had a predictive map stored in his brain. He pointed out non-descript yards and houses where he had been called to the scene of homicides, burglaries, and gang shootings.

He continued through particular blocks of residential neighborhoods where HunchLab had placed red and orange boxes to indicate a risk of aggravated assault and gun crime. The streets were lined with crumbling little brick houses. In a few yards, signs reading "We Must Stop Killing Each Other" had been stuck into the dirt.

As Keener drove through the bottom left corner of a HunchLab box—red to indicate a high risk of aggravated assaults—he noticed a white Chevy Impala with a dark window tint, dark enough to merit a traffic ticket.

Keener gunned his motor and flashed his lights. The car slowed to a stop, and Keener walked up to the window. Leaning down, he caught a whiff of marijuana. The young man was black and looked to be in his 20s, with a baseball cap, grey sweatpants, and a tattoo that crept out of his shirt. Keener said, politely but firmly, "I smell what smells like weed to me."

The man said he smoked earlier, but that there was "nothing in the car."

Keener decided the smell gave him probable cause for a search. He told the young man to step out, frisked him, and asked him if he had anything "I should know about." The man said he had a gun. Keener found a black Glock 23 pistol, .40 caliber, under the seat. He took it back to his car, noticing it had no magazine—just a bullet in the pipe. "Pretty big," Keener said, turning the gun in his hands.

Another police car rolled up behind Keener—a standard call for backup had gone out. Standing between his own car and Keener's car, the young man stared at the ground, clearly annoyed but also trying not to appear annoyed. When Keener asked about the smell again, he said, "You ask me if I smoke. I smoke, man!"

While the man paced, Keener looked up his name and the gun's ID number. It didn't turn up as stolen. It is legal to have a gun in your car in Missouri if you're over 18 and not a convicted felon. The man had been arrested, but never convicted, for stealing a gun. Keener let him go with a ticket for the window tint.

As the Impala drove off, Keener looked back at the HunchLab map. The stop itself had gone down just outside the aggravated assault square. "He could have been going to shoot somebody," he said, shrugging. "Or not."

That HunchLab had sent him to a location where he may or may not have averted illegal activity was, for the moment, tangential; it would not be clear for months whether crime rates in Jennings might be affected by the program.

Research on the impact of predictive policing programs is still in its infancy. Last year, PredPol researchers published a study finding that sending patrol officers to several areas of Los Angeles predicted by their algorithm led to a reduction, on average, of more than four crimes per week in these neighborhoods—twice as efficient as human crime analysts. The researchers said the savings—resulting from not having to investigate and prosecute crimes that otherwise may have happened—could reach $9 million per year.

Jeremy Heffner, the product manager for HunchLab, is careful about making promises; he argues the results will vary based on how a particular police agency uses their analysis. "By having more accurate locations, we amplify the effect a meaningful tactic may have," Heffner told me, "but you still need a meaningful tactic." Studies of HunchLab's effectiveness are underway in several cities, and researchers in Philadelphia are comparing patrolling in marked police cars to sending unmarked cars, which could quickly respond to crime, but might not deter it.

Even with data-driven tools, on-the-ground police work is full of ambiguity and discretion, which makes measuring their impact difficult. Would Keener have stopped the car at all had it not been in a HunchLab box? "It's all relative," Keener said. "Probably." He was careful to point out that being in the box alone was not a good enough reason to stop someone. "Does the data give me grounds to stop just because they're walking around? No."

"My son don't have anything positive to say, so he'd rather not say anything at all," said the mother of the man whom Keener had stopped. It was a few weeks later, and we were talking by phone—her son had not wanted to be interviewed, and she was too suspicious of the police to put her name in print. "Believe it or not, if you say anything to the press," she said, the police "will make sure to pull you over and treat you worse."

The mother and her son have been pulled over a lot, she said, and it often feels as though they are targeted because they're black. "They give you a reason"—the tinted windows, the marijuana smell—"but then they get to asking you to get out [of the car]. Well, why do I have to get out? Because you said so? All I can go on is it's because I'm black."

There are widespread fears among civil liberties advocates that predictive policing will actually worsen relations between police departments and black communities. "It's a vicious cycle," said John Chasnoff, program director of the ACLU chapter for Eastern Missouri. "The police say, 'We've gotta send more guys to North County,' [where Jennings is located] because there have been more arrests there, and then you end up with even more arrests, compounding the racial problem."

Dolly had tried to mitigate this issue by having HunchLab identify two boxes for every beat, no matter how "high-crime" the area. But to Chasnoff, the entire emphasis on patrolling was misplaced. "I don't think anyone, in the abstract, has a problem with figuring out where crime is and responding to it," he said. "But what's the appropriate response? The assumption is: we predicted crime here, and you send in police. But what if you used that data and sent in resources?"

Many at the top of the county police agree with Chasnoff, and are interested in finding ways to use predictive policing software to address crime through other government resources. "We can't just have the criminal justice system solve our problems," Belmar, the St. Louis County police chief, said the morning before Keener's shift.

Late last year, the department began sending lists of high-crime areas to a nonprofit called Better Family Life, which deploys outreach workers to connect residents with drug treatment and educational programs. In theory, HunchLab could provide even more targeted areas for this organization and others to apply their model of what James Clark, the nonprofit's vice president, calls "hot-spot resources."

This has not happened yet. For those who would like to see police change their methods in the wake of Ferguson—and shift toward problem-solving and community relationships rather than patrolling—HunchLab looks like one method of data analysis being swapped for another. It may be more objective and may lead well-meaning police commanders to be more thoughtful about what's driving crime, but that's little comfort to the young black men stopped and searched. "It's another example of the county police selling themselves as more professional," Chasnoff said, "but maybe it's just a more professional use of the same bad ideas."

Throughout his shift, Officer Keener witnessed hints of simmering distrust. At one point, several children danced in the street, which he said was locally understood to be an insult to police. Later on, just outside one of the HunchLab boxes, he drove by a house where, he said, a drug dealer lived. A suspicious-looking rental car was parked outside. As he slowed down and peered out, the car door opened and a woman—black, maybe mid-30s—emerged. She pointed at the police vehicle to someone in the car and scrunched up her face in disgust. Keener turned back toward the dashboard screen and rolled away.

DONALD TRUMP USHERS IN THE ANTI-FUTURE AGE

Hal Niedzviecki

(First appeared in The Dark Mountain Project, *December 20, 2016)*

During the 2016 presidential campaign in the USA, and during the primary season which preceded it, the eventual winner Donald Trump was the only candidate to come out as hostile to that darling of both traditional parties and their corporate pals, the technology sector. Asked about his thoughts on the rise of Artificial Intelligence (AI) in an interview, he said: "I have always been concerned about the social breakdown of our culture caused by technology. I think the increased dependence and addiction to electronic devices is unhealthy."

When US high-tech industry CEOs started a lobbying and education group to advance the cause of expanding the H-1B visa program, which they rely on to bring foreign computer programmers and engineers into the country, Trump was unmoved. In fact, one of his few clear policy pronouncements was to oppose the H-1B visa program altogether. Trump also said he would initiate anti-trust action against Amazon and promised to force Apple to make its products in the United States—later calling for a boycott of the company when it declined to unlock the iPhone of a suspected terrorist. And sure, Trump used social media in his campaigns, but not the way the other candidates did. As one commentator noted, candidates were "dropping $5,000 to $10,000 per month" on social media management and delivery systems like Sprinklr or Hootsuite, "with the exception of Mr. Trump who definitely" managed "all his accounts by hand."

While Trump was trashing or ignoring the techno standard bearers, his opponents were eagerly deploying slick social media and analytics technologies and kowtowing to the new gods of so-called innovation. BlackBerry-bearing Jeb Bush hailed an Uber to the San Francisco startup Thumbtack where he praised the digital economy to the skies. Hillary Clinton traveled to San Jose, CA, in Santa Clara County, the center of Silicon Valley, national leader in average wages despite a poverty rate higher than 10 percent. Clinton called San Jose "a city that's all about the future: the future of our economy, the future of our society, of how we're all going to be stronger together." During his presidency Barack Obama visited an Amazon factory and compared the workers there to Santa's Elves, completely ignoring Amazon's reputation for mistreating perpetually precarious workers, shuttering stores on Main Street and soaking up state tax breaks for the privilege; his White House website featured a whole section dedicated to innovation and "winning the race to the future." Back when Trump was perceived as just an annoyance, Hillary Clinton was giving speeches talking about the "on demand or so called 'gig' economy...creating exciting opportunities and unleashing innovation."

Trump rejected the prevailing ideology of future-first both in words and in symbol. His campaign slogan—Make America Great Again—evoked not the future or even the present, but the past. Trump conjured up a fantasy era when America was (mostly) white and mostly just one union job on the assembly line away from a car and a house and a wife who stayed home and took care of the kids. His signature campaign policy was also very instructional: a wall separating the US from Mexico. A wall. Not lasers or satellites or big data or colonizing Mars or getting to the future first. Trump constantly talked about his real estate holdings, his wealth. During the primaries and presidential campaign, he took time out to cut the ribbon on a new Trump hotel in Washington DC. He visited his Scottish golf course. He held forth on his net worth and his ability to outfox creditors and evade the taxman. Trump embodied his rhetoric—make real things, cut through the crap, get rich and get away with whatever you can.

These thoughts whirled through my mind throughout the campaign. I found myself fascinated with the highly ironic fact that of all the candidates, only Trump seemed to be able to signal (if not actually say) that, as I and others have argued, many of the so-called advances in technology over the last twenty years have been harnessed to create efficiencies that significantly reduce employment and compensation. Only Trump was able to repeatedly signal agreement with the conviction that, as one journalist wrote in the aftermath of the election, Silicon Valley "is out of step with [a] national and global mood" beset with "social and economic anxieties" at least partly caused "by the products the industry devises." For Trump, this was just a part of his overall bricks-and-mortar-build-the-wall rhetoric; but it was a terribly important part, a core element of his spiel's appeal.

Still, though Trump, as with Bernie Sanders, sensed the intense anxiety around technological displacement and precarious labor, though he seized on it, he didn't—and doesn't—have any meaningful course of action to resist or rebut it. It's not like he was out garnering votes in small-town Ohio by telling the unemployed that the virtual Eden perpetually promised by the demi-gods of future is a dangerous chimera that needs to be replaced by pragmatic action to replace consumption with community. As *New York Times* columnist David Leonhardt writes: "Clinton didn't lose to Donald Trump because he had a more serious set of policies for revitalizing working-class America." Trump's election, like his anti-future campaign, is a rebuttal, but not an answer.

Instead we should think of Trump's win as the sudden swoon from flight into lifelessness for the canary in the coalmine. It's as evocative a sign as the relentless California drought and the looming extinction of most of the planet's largest mammals. The ascendance of Trump signifies that we've moved past the point of conventional politics, past, in some ways, the point of return. It signifies, if nothing else, that progress is hard to sell in a world that is literally getting hotter and more unlivable every year. Though many reject man-made climate change (a Chinese conspiracy, blusters Trump), and many more of us refuse to give up the luxuries that might avert the worst, the spiritual emptiness of living, powerless, in a time of perceptible decline, is our primary common value. We all find ourselves emotionally emptied with noticeably fewer prospects than our parents and even their parents.

In response, more and more of us are rightly rejecting technology-as-savior and future neoliberalism (from both the right and left). But with no better option being offered to us, into the void the strongman is promising to return us to ancient ritual, the sword and the scythe, blood dripping onto land. With Trumpites pumping nostalgia direct to our veins, far-right conspiracy theory preppers round the circle with far-left anti-vax dropouts. This is what a Trump win signifies: the post-political void at the end of the capitalist fantasy. Guess what? The world was burning while we looked for meaning in our Netflix echo chamber of *Stranger Things* and *Celebrity Apprentice* reruns, neato plans to terraform other planets and social media campaigns to save the species of the month.

Like a lot of people, I stuck my head in the sand during the campaign and said to myself: not yet; it's too soon; I'm with her. I second-guessed myself, talked myself into believing that we hadn't yet gotten to the point at which all we had was rejection, was void. But like everyone else, I was wrong. It happened. It is happening. Into the vacuum of contemporary politics voided by technological displacement, rapacious capital and the existential threat of climate change arrives Donald Trump, Brexit, Putin, Turkey's Erdoğan and the Philippines' Rodrigo Duterte—anti-future emblems of an unquenchable thirst for simplified belief systems, circles of life, the easy path back to the idyllic pre-literate past.

These yearnings are not solutions or ends. They are, instead, impulses and desires, angry fires burning our brittle principles right down to their foundations. In the anti-future we build walls. We fall back upon primal impulses. We turn inward, turn tribal. We rage impotently against the machine's virtual reality rainbow and seek an impossible return to a real which doesn't exist, which never existed. We vote Trump for president because, if nothing else, he was the only one; the only one who told us he loved us; the only one who promised to save us from the future and make us great again.

THE BATTLE FOR THE GREAT APES: INSIDE THE FIGHT FOR NON-HUMAN RIGHTS

George Johnson

(First appeared in Pacific Standard, *November 21, 2016)*

On a hot summer day at the end of January, I walked up Calle Borges from Plaza Cortázar (surrealist writers are honored in these parts) for my first visit to the Buenos Aires Zoo. As a boy, Borges himself came this way to gaze through the bars at the Bengal tiger—a beast so fantastic that it lodged itself in his imagination, haunting him ever after in his words and dreams.

I passed quickly by the pony rides and the man selling balloons at the wrought-iron entryway, and past Lago Darwin, where pink flamingos swam. Past the dark reptilario, torpid with snakes and lizards. Past the bears and the lions and the cheetah and the ocelot. And finally, just beyond the towering condor cage, I saw her: red-headed Sandra, who, in the eyes of some humans, has become the world's most important ape.

Last year, to the delight of animal-rights advocates, a Buenos Aires judge ruled that Sandra is a "non-human person" and a "sentient being"—a bearer of legal rights. Just what that means is still a matter of dispute. But, in the meantime, the case has been hailed by activists as a milestone in civil rights—another sign that human society may be ready to expand its embrace, recognizing great apes and perhaps other species as more than just things.

Standing before the glass wall of Sandra's enclosure, I looked across an empty moat into her open-air habitat—one of those *Flintstones*-like rockscapes that in the 1960s began replacing barred cages in the architecture of zoos. There she was, sitting alone in the shade of an artificial cliff, hiding beneath her blanket. Unlike the screaming chimpanzees nearby, orangutans are quiet creatures that guard their privacy. This was an animal I could identify with.

While waiting for her to show her face, I tried to decipher the educational signage. Orangutans, which live in reduced numbers in Sumatra and Borneo, are known by the natives as *hombres del bosque*—"men of the forest." Threatened with extinction, their hope for survival lies in *conservación* and *sustentable de su hábitat*. The sign, like almost everything else at the zoo, was sponsored by Coca-Cola.

As I puzzled over the words, I glimpsed from the corner of my eye a flicker of movement. Sandra? But it was just a reflection of a boy who had come with his family to see the orangutan. They quickly grew bored and moved on.

It wasn't until the next afternoon that I got a better look. Sandra had returned to the same spot, but now her head was sticking out from the blanket as she scrutinized the ground for insects, popping them like candy into her mouth.

A group of visitors approached, guided by a young woman wearing khaki shorts and an official zoo polo shirt. Sandra recognized her. She rose from the ground and ambled, as if in slow motion, across her paddock and into the building that served as her indoor quarters. Looking out through the glass, she matched hands with her human keeper, mano a mano, and I wondered, *What was Sandra thinking?* Was this a mindless act of imitation? A sign of affection? Or could it be an entreaty? *Why do you taller apes get to be out there with the towering trees, where space appears to be infinite?*

Then, as if remembering her part of the contract, Sandra walked back outside and climbed slowly onto a platform built to provide her with a semblance of arboreal existence. After a round of acrobatics, she returned to her corner to pick for more bugs.

We hold these truths to be self-evident, that all men are created equal, that they are endowed by their Creator with certain unalienable Rights. In 1776, when the Declaration of Independence shook the world, "men" meant men, and Thomas Jefferson, the man who drafted these words, counted among his possessions hundreds of slaves.

But society moved on. First came abolition and emancipation. It wasn't until 1920—less than a century ago—that women in the United States secured the right to vote. Now there are foreshadowings of what some see as the next logical and moral step.

Feld Entertainment, whose Ringling Bros. and Barnum & Bailey Circus has been a target of animal-rights activists, recently decided to retire the last of its traveling elephants to a 200- acre conservation center the company operates in central Florida. Not long after, SeaWorld ended its breeding of orcas as part of a plan to replace killer-whale shows with what the company called more "natural" encounters.

But the pressure for reform has been especially intense with regard to great apes, creatures for which—or for whom—humans feel the strongest kinship. In May, when zookeepers in Cincinnati shot dead a gorilla to save a child who had fallen into its enclosure, waves of outrage were aimed at the marksman and the boy's mother, whose negligence was treated as the equivalent of manslaughter.

Bowing to sentiments like these, the National Institutes of Health began phasing out chimpanzee research in 2013. Most experiments critical to human health, an advisory panel recommended, could be performed as effectively on other animals. Last year, captive chimpanzees were added to the list of endangered species maintained by the U.S. Fish and Wildlife Service (wild ones had already been included), and Francis Collins, the NIH's director, announced that most of the institutes' chimps would be retired to sanctuaries—a migration that is already underway.

But with Sandra the orangutan—and Cecilia, a chimpanzee in another Argentine zoo who is also under consideration by the courts as a non-human bearer of rights—the stakes are higher. No one is proposing that they be allowed to vote in an election or run for office, simply that they be recognized as thinking, feeling beings in a court of law. In the U.S., the Nonhuman Rights Project has been pursuing a similar kind of recognition for Hercules and Leo, two recently decommissioned laboratory chimps, and for Kiko and Tommy, another pair of chimps. Kiko is being kept as a pet, and Tommy's last reported whereabouts was a zoo in northern Michigan. But with Sandra, Argentina has moved to the forefront.

Animal-welfare laws have long held that we should avoid making other creatures suffer—to the extent that this doesn't interfere unreasonably with human commerce. Many people think that strengthening these ordinances is enough. Others look to commerce for a solution. Wayne Pacelle, the president of the Humane Society, has argued that the animal-products industry, responding to pressure from consumers, has been steadily improving conditions. His new book is called *The Humane Economy*. But some animal-rights lawyers are skeptical that moral decisions should be left to the marketplace. They are calling on *Homo sapiens*, the self-declared wise ones, to go further, enlarging the definition of what counts as a person—a legal entity entitled to certain rights.

Persons, as defined by the law, are not necessarily human beings. Corporations, after all, have long been considered "juridical" or "artificial" persons with some of the rights and obligations of people. They can sue and be sued and have rights to freedom of speech (as in *Citizens United*) and even, to a more limited extent, religious expression (as in the *Hobby Lobby* case).

Off in another realm, legal scholars foresee a day when a judge will decide whether to grant personhood to an artificially intelligent computer program or to a chimera created in a lab by fusing human and non-human genes.

So where does Sandra—and all the world's great apes—fit into this jumble: species just a twig over from *Homo sapiens* on the tree of life, but without the capacity to argue on their own behalf?

It's been more than half a century since Jane Goodall first filmed chimpanzees in Tanzania fishing for termites with sticks they had modified by stripping off the leaves—a simple example of making and using tools. Chimps have also been observed using a pair of rocks, a big one and a small one, as a hammer and anvil for cracking nuts—and even adding a third rock as a wedge to stabilize their platform. Chipped pieces of granite that archeologists suggest are chimpanzee nutcrackers have been discovered in Ivory Coast and radiocarbon dated to 4,300 years ago.

Other apes show a similar aptitude for gadgetry, and orangutans seem to have devised an especially neat trick: They put sticks in their mouths, lowering the pitch of their voices and making themselves sound bigger and scarier to potential enemies.

Toolmaking itself doesn't seem like grounds for a constitutional amendment. If so, we would also have to include New Caledonian crows, which make probes and hooks from twigs and leaves for retrieving elusive grubs.

Self-awareness may be a more persuasive criterion. Great apes, many primatologists argue, can recognize themselves in mirrors. They exhibit signs of altruism and grieving for the dead, and they appear to know that other creatures have desires and intentions—what psychologists call a "theory of mind." That comes about as close as you can get to showing that another creature is a vessel of consciousness and an autonomous agent possessing something like free will.

But these abilities too, in various dilutions, have been demonstrated elsewhere in the animal kingdom. Set the bar too low and you end up emulating the Jains in India, wearing masks to keep from accidentally inhaling and killing flies.

What about an aptitude for counting and other crude arithmetical skills? Those have been reported in great apes—and also in parrots,

"What we're left with is a hierarchy of intelligence—and, presumably, consciousness. And it's we humans who have the power to arbitrarily draw the line."

newborn chicks, and lizards. There is a small body of literature claiming that ants follow Fermat's Principle of Least Time to find the shortest path to a source of food (which is a little like arguing that planets solve differential equations as they orbit the sun).

In all of these cases one must be wary of anthropomorphism. But just as perilous is what the primatologist Frans de Waal calls anthropodenial: the mystical belief that consciousness sprang into existence only with the birth of humans.

The most persuasive evidence for human-like intelligence may be an ability to communicate with sounds and symbols, including hand gestures and lexigrams on a keyboard. We've all heard about the linguistic feats of Kanzi the bonobo and Koko the gorilla. But Alex, the grey parrot trained by Irene Pepperberg, a well-known researcher in comparative psychology, also showed uncanny communication skills.

All of these claims are subjective and open to dispute. Maybe genetic similarity is a better measure. Great apes, it's commonly said, share as much as 99 percent of our DNA. (The precise number depends on how you do the counting.) But for mice the overlap is still as high as 90 percent. And there are those eerily intelligent whales, porpoises, and dolphins way off on their own limb of the evolutionary tree.

In the end, the science seems almost beside the point. Brains are organs that think, and any creature with a brain—or even a ganglion—processes information. What we're left with is a hierarchy of intelligence—and, presumably, consciousness. And it's we humans who have the power to arbitrarily draw the line.

The planet's alpha apes, not Jefferson's "Creator," are the endowers of inalienable rights. And that's where the politics comes in—people with conflicting values and assumptions competing to do what they think is right for the world.

Late one afternoon, shortly after my visit to the zoo, I arrived at a law office on a jammed, noisy block of downtown Buenos Aires to meet with a group of activists eager to explain, with the help of a translator, why they believe humans are not the only creatures deserving of legal rights.

In my ignorance of the language, their words came in a rush—layers of Spanish colliding with English that would later take hours to sort out. I wondered if this was how Sandra feels as she huddles beneath her blanket and listens to the clash of human voices.

"Animals are not objects to satisfy outside needs," said Gerardo Biglia, an animal-rights lawyer who filed a friend-of-the-court brief in the Sandra case. "What most interests us is that zoos not exist."

As she prepared a gourd of *mate*, Malala Fontan explained how their group—called SinZoo ("Without Zoo")—came together in 2014 after the city announced its annual Night of Museums. The zoo, along with other public institutions, would be kept open after dark for people to enjoy the exhibits. It wasn't just the additional stress on the animals the group objected to, but the idea that they were being treated like paintings and sculptures.

"Any animal in the zoo is suffering from being captive and from being exhibited," Fontan said. "This is like slavery."

The group also objected to the city's earlier decision to turn over operation of the zoo to a private subcontractor—another reminder that the animals were considered not just inmates but commodities.

Aldo Giudice, a biologist who had joined the conversation, proposed that the effects of imprisonment are especially hard on intelligent creatures like great apes, which can see people coming and going without constraint. Called on to submit a friend-of-the-court brief in the Sandra case, he spent many days observing her.

"Sandra is bored," he said. "She lies there waiting for stimulation from her keepers." Deprived of trees to swing through, he believes, she has become depressed, overweight, and anxious. "It's like being locked in an elevator your whole life, 365 days a year."

I described how I had seen her, the day before, putting her hand to the glass, reaching to touch the hand of a young woman.

"In all jails you have interaction with the jailers," Giudice said. "Sandra is lost in the heart of a perverse system that treats her like an object—like a camel, a horse. All the animals get treated the same, except that Sandra is more like us. Like a human."

"I have heard nearly as much nonsense about zoos as I have about God and religion," says Piscine Molitor Patel, the son of a zookeeper in *The Life of Pi*, Yann Martel's novel about the détente we strike with the other creatures of the world. "The life of the wild animal is simple, noble and meaningful....Then it is captured by wicked men and thrown into tiny jails. Its 'happiness' is dashed. It yearns mightily for 'freedom' and does all it can to escape. This is not the way that it is."

It is the wild, in Pi's estimation, that is the prison with the constant worries about finding food or being eaten—the struggle to survive. In such a world what can be the meaning of freedom? Living in a zoo is like checking into a hotel.

I was thinking of that passage as I sat in the small office of Adrián Sestelo, the Buenos Aires Zoo biologist—a man who clearly cares about Sandra and all of the animals in his charge but who is frustrated by the romantic notions society thrusts upon them.

"The people who work in the zoo are not evil people who are keeping animals captive for their own enjoyment," he said. "We are proud professionals who are concerned about conservation."

He told of a jaguar run down by a car in the northern part of the country. The body was sent to his lab so that the animal's sperm could be cryogenically preserved and used to increase the diversity of the zoo population. Beyond firing the imaginations of children, the animals in the zoo can serve as a repository for this genetic information—DNA that can be introduced back into the wild to help ensure the survival of a species. There are reasons for keeping animals other than exhibition.

He acknowledged that conditions at the zoo, which had opened in 1875, needed improvement and that its mission must change. Plans were already in the works, he said, to move away from displaying exotic creatures, concentrating instead on local fauna and teaching visitors about the importance of protecting wild habitat. For that reason he is not opposed to sending Sandra to a sanctuary or a zoo where there are other orangutans. But it is just as important, he stressed, that there be people around her. "She needs contact with humans," he said. "This is normal for her."

Sestelo is dubious that great apes are especially deserving of something called personhood. "I don't think it's right to draw some animals closer to our laws and keep other animals more distant," he said. "It takes us back to medieval days when man was considered closest to God."

"It's not fair," Sestelo continued. "Why not the cow, the pig, the chicken?...It is almost racist." All species, he said, have a right to be on this planet and to live in their unique way.

As our conversation drifted into the philosophical—the legacy of Descartes and A.O. Lovejoy's *The Great Chain of Being*—I asked if he thinks Sandra is happy.

It was a question he had heard too many times before. "I can't tell you," he said with a hint of exasperation. "That is a human feeling. People talk about Sandra's sadness without understanding the biology of apes or orangutans. It's easy to wonder if they are conscious because they are similar to us. But why not ask the same question about spiders?"

Deep in a sub-basement of Western jurisprudence is an assumption that has rarely been questioned: that there are precisely two kinds of entities under the law, persons and things. A comatose human kept alive by machines holds rights not accorded to the smartest ape. Last year, Judge Elena Liberatori challenged the dichotomy, declaring that Sandra is a non-human person with fewer rights than people but more than, say, a lump of clay.

The case began in November 2014 when a group called AFADA (the Spanish acronym for the Association of Officials and Lawyers for Animal Rights) petitioned a Buenos Aires court for a writ of habeas corpus. Invoked by criminal defense lawyers on behalf of their clients, habeas corpus is a demand that a prisoner be brought before a tribunal to determine whether his or her confinement is justified. The maneuver, which dates from medieval times, has been adopted as a tactic by animal-rights advocates. A judge ordering habeas corpus for an ape would be implicitly recognizing the animal's personhood.

To no one's surprise, the petition for Sandra was quickly dismissed—and dismissed again on appeal. But then a federal appeals court took a more ambiguous stance. The three judges, Alejandro Slokar, Ángela Ledesma, and Pedro David, also declined habeas corpus, remanding the case to a lower court to decide whether animal cruelty had been committed. But in their order the judges referred, in passing, to Sandra as a non-human holder of rights.

Seizing on those words, AFADA filed a new case—this one in a Buenos Aires court where citizens can seek redress against the government. And so it fell to Liberatori to consider the simian's fate.

It is hard to imagine that Sandra could have been dealt a more sympathetic judge. When Liberatori met me in her office near the Plaza de Mayo, where Eva Perón once stirred the masses, she was wearing striped pants and tennis shoes. On the wall was a cartoon showing her dressed in rags like a homeless person and declaring, "We have to urbanize."

"The slum judge," the caption called her. The reference was to an incident in which she ordered the city government to extend services to one of Buenos Aires' shantytowns. She was also the first judge in Argentina to rule that a marriage ceremony between two women could take place. (And the second to do so for two men.) When the Sandra case came to her in March 2015, she was prepared to make more waves.

Working with a young anthropologist, Lucia Guaimas, she solicited advice from orangutan experts in Argentina, the U.S., and Australia. She also dug into the literature of animal-rights law, including a book, *La Pachamama y el Humano* (*Mother Earth and Humans*), by Eugenio Raúl Zaffaroni, a former justice of the Argentine

Supreme Court who is now a member of the Inter- American Court of Human Rights.

I asked if he was an influence on her thinking. "*Totalmente,*" she said.

In October she handed down her decision, a remarkable document that reads in parts like a dissertation in postmodern philosophy. "All forms of classifying and categorizing the world are a social construction," Liberatori writes, sounding the jargon. That includes our conventions about "what is considered superior and what is considered inferior and who or what should have rights."

Marshaling reasons why it may not be so crazy to consider Sandra a "non-human person" and "rights-bearing subject" (like you, me, and Monsanto), Liberatori draws on a wide range of sources. Recently, a revision of France's Napoleonic code (as influential on Argentina's civil code as English common law is for the U.S.) added "sentient beings" as a legal category, a move already taken by some other countries. And the constitution of Ecuador, she noted with a nod to Zaffaroni, recognizes nature itself as a bearer of rights.

Looking closer to home, the judge considered instances in which the Argentine legal system already appeared to look upon other creatures as subjects rather than objects. A federal anti-cruelty law refers to animals as "victims," suggesting a right to be treated with dignity and respect. When sniffer dogs used by federal customs agents reach retirement, she noted, the state provides them with housing, health care, and food—something you wouldn't do for a surplus truck.

"There is no doubt that even if a living being's life and dignity are completely outside the realm of the legal system," she wrote, "this does not stop them from being extended analogically from 'human people' to Sandra when she is granted the condition of 'sentient being.'"

After many pages of such considerations, she arrived at her conclusion: "Sandra has the right to enjoy the best quality of life possible, tailored to her individual circumstances."

"To achieve this," Liberatori wrote, "we must tend toward avoiding any type of suffering that could be generated by Man's meddling in her life."

Acknowledging that the orangutan was born in captivity and had never known the wild, Liberatori appointed a panel of advisors to determine whether Sandra should be freed to a sanctuary or remain under improved conditions at the zoo. In the meantime she ordered Sandra's keepers to take steps "to preserve her cognitive abilities."

For the plaintiffs, it was a landmark moment. "There has never been a decision of this magnitude," said Andrés Gil Domínguez, AFADA's lawyer. "It was historic, a turning point in animal rights." Not long afterward the case was cited by another Buenos Aires court, which ruled that 68 mistreated dogs were, like Sandra, rights-bearing subjects. More decisions like this are bound to follow. Maybe one will go as far as granting habeas corpus.

As expected, the city appealed—as did AFADA, which was disappointed that the judge stopped short of deciding Sandra's fate, deferring instead to an expert panel.

In June, eight months after Liberatori's decision, the Court of Appeals of the City of Buenos Aires handed down its ruling. The judges agreed that there was no need for the advisory panel. Moreover, they upheld Liberatori's demand that the city provide Sandra with more-stimulating surroundings, taking into account her "well-being, behavioral complexity, and emotional states." Later, if it seemed in the ape's best interest, the zoo could send her to a sanctuary.

As for the larger question of non-human rights, the judges decided to leave that open for future tribunals to decide. The matter, they wrote, "*no es pacífica*"—it remains unsettled, a controversy still unfolding. They disagreed with Liberatori that federal law, as it currently stands, implies that the zoo must treat Sandra as a rights-holder. But whatever one's position on non-human personhood, they observed, "Nobody now questions that the suffering of animals must be outlawed and that humans have a duty to care for them."

Later that month, animal-rights advocates claimed another victory. Declaring that the zoo "generates more sadness than happiness," the mayor of Buenos Aires announced that the city was closing the gates and accelerating plans to create an *ecoparque*, much along the lines Adrián Sestelo, the zoo biologist, had described to me. Most of the 2,500 animals will gradually be moved to nature reserves, while those too old or infirm will be cared for onsite but no longer displayed to the public.

Sandra would remain, at least for the time being, wondering perhaps where all the people went.

Before flying to Buenos Aires, I had downloaded the original *Planet of the Apes* to watch on the plane. I'd forgotten that the orangutans in the story are actually bad guys—evil scientists and upholders of orthodoxy who happily perform brain-ablation experiments on dumb humans. The gorillas are even worse, while the heroes are the chimpanzees, Zira and Cornelius—upholders of non-simian personhood and human rights.

Six hundred miles across the Argentine pampas, in Mendoza, a court is considering the fate of a chimpanzee named Cecilia—the target of another habeas corpus suit by AFADA.

It's easy to say when you're on this side of the bars, but I liked the Buenos Aires Zoo, with its arching tipa trees and Victorian architecture—nature merging harmoniously with artifice. I'm glad I got to see it before it closed. The Mendoza Zoological Park was a mess. Reports of dying animals had been in the local news in recent months. On the day I arrived the zoo had just reopened temporarily after being closed for safety reasons by its newly appointed director, Mariana Caram. (Three months later it was closed again, this time indefinitely.)

She was a most unusual choice for the position. A former Fulbright Scholar working in sustainable development, Caram had been collaborating with environmental organizations pushing to reform the zoo. Hired last fall by a newly elected provincial government, she is trying to clean up the place and eventually turn it into an ecopark like the one now under way in Buenos Aires.

She must deal, in the meantime, with the matter of Cecilia, who became a poster chimp for activists after an incident in 2014. A pack of wild dogs had charged through the grounds and killed 27 rheas, a vicuña, four guanacos, and two llamas. During the rampage, one of Cecilia's companions, a chimpanzee named Charly, died of a heart attack, and, six months later, another chimp, Xuxa, succumbed to what were said to be natural causes.

With Charly and Xuxa gone, Cecilia, the youngest, was left on her own in a concrete habitat, with no vegetation or natural ground to walk on—"one of the most horrible enclosures in the zoo," Caram said. She hopes the court rules in favor of recognizing Cecilia as a non-human person. In the meantime, discussions are under way about moving her to a sanctuary in Sorocaba, Brazil.

"I know I'm crazy taking on the zoo," Caram said as we walked from her office to a bear enclosure with a fetid pond that smelled like sewage. Just up the hill, kept out of sight from visitors, was a jail-like building with two more bears. There was no other place to put them. "We talked about non-human persons," she said, "but do we have the right to keep any animal like this for its entire life?"

Farther along the path was the worst cage of all, packed with more than 100 frantic baboons. "No birth control," Caram said. Two weeks earlier some of them had escaped, and one injured a girl.

"Look at this—concrete, concrete, concrete," Caram said as we walked quickly past more cages. "Why does an animal have to live like this?"

We ended with a visit to Cecilia and one of her keepers, Daniel Garrido—a third-generation zoo employee who has cared for the chimpanzee since she was born. He was worried about how she would fare at a sanctuary with strange new chimps. When Charly and Xuxa were alive, he said, they would gang up on her. Though she was depressed temporarily by their absence, he has since seen her thrive. Animals, like people, are individuals. Maybe she prefers being on her own.

You could tell he would miss her.

"Are you a father?" he asked. "It would be like taking away a child."

As the campaign for non-human personhood presses on, hundreds of chimpanzees in the U.S. are being retired from medical research and moved to primate sanctuaries like Save the Chimps in Fort Pierce, Florida.

Carved out of old citrus groves, this simian paradise has grown to include 12 man-made islands and 250 chimpanzees. Each island is connected by a land bridge to a building on the mainland where the chimps are fed and receive veterinary care. If they choose, they can spend the night inside.

They are still things under the law. But they are allowed to live beyond the gaze of the public, and with some of the autonomy that seems appropriate for sentient beings.

One of the first to arrive was Cheetah, who boarded a trailer in Alamogordo, New Mexico, five years ago to join what became known as the Great Chimpanzee Migration. Leaving behind a biomedical lab called the Coulston Foundation, he and nine companions, each with a window seat, began the two-day journey to Florida.

"Cheetah endured multiple liver biopsies, possibly without anesthesia," Molly Polidoroff, the sanctuary's executive director, told me when I arrived there early this year. "That he can be so people-friendly just amazes me. We have other chimps that, not surprisingly, will throw things. And you kind of can't blame them. We're not their friends in a lot of cases."

As we toured the grounds in a golf cart, she pointed across a lagoon to Air Force Island, where the sanctuary's first residents, retired from the U.S. space program, arrived in 1997. Stopping across from Alice's Island (named for one of its inhabitants), she introduced me to Cheetah.

Reaching through the mesh with a piece of rubber hose (a "tickle stick"), he tried to untie my shoelace while his friend Timmy competed for attention. Nearby, a group of chimps played with torn colored paper left over from a "Chimpmas" party.

At the outbuildings across from another island, Jocelyn Bezner, the sanctuary veterinarian, told me about Bobby. "He was definitely the most traumatic," she said. When she first saw him back at Coulston, living in a place his rescuers called the Dungeon, his left arm was like a chewed piece of meat.

"Every day he would scream and bite a chunk out of it," she said. Somehow his nerve fibers had become damaged and were sending spurious signals to his brain. Several surgeries later, his suffering was relieved and he was introduced into a group of other chimps. He rose to become leader of what is now Bobby's Island. He was like Gandhi, Bezner said, breaking up fights between other chimps. Later, when another male challenged him as alpha, he gracefully stepped aside.

Before the day was over I saw Clay, taken from his mother hours after birth and used to test the toxicity of ibuprofen and other pharmaceuticals. And there was Rebel, born at Coulston and then rented to a lab at the National Institutes of Health.

And I met Lisa Marie, the newest resident, who had just turned nine. She had been rescued from the entertainment industry—she'd been owned by an Elvis impersonator who used her in animal shows. Still afraid to venture onto her new island home, she was living ashore and slowly being initiated into chimpanzee society. When a chimp named Christopher intimidated Lisa Marie, Bezner brought in two others whom she knew Christopher would defer to. I was struck by how the chimps—they all looked the same to me—were treated as individuals. Through trial and error, the right family would be found for Lisa Marie.

I wished that Daniel Garrido, the zookeeper in Mendoza, could see this. It might help relieve his worries about Cecilia's going away.

After leaving Fort Pierce I drove inland past Yeehaw Junction and signs advertising "P-Nuts and Gator Jerky" to Wauchula, where a Family Dollar sat next to a Dollar General and across the street from a Dollar Tree—products of our superior human intellect. A few miles outside of town I found the Center for Great Apes, the only sanctuary in North America licensed to keep orangutans as well as chimps.

Shaded by a canopy of dense forest, the center is home to 30 chimps and 15 orangutans living in a dozen spacious enclosures, each about three stories high and connected by a mile and a half of elevated chutes. Taken from their mothers as infants, most of the apes worked in Hollywood or in roadside zoos, or were kept as pets. Once they became too willful, their owners didn't want them anymore.

"At two years old they're stronger than we are," Patti Ragan, the founder and director of the center, told me. She started the sanctuary 23 years ago, when she was trying to find homes for an infant orangutan and chimpanzee who were being kept at a Miami tourist attraction. At around age five, young apes begin to have tantrums, she said, and after seven or eight years they've passed their "shelf life." But they are still youngsters at that point, with as many as 40 more years to go.

As we walked past the domed structures, Ragan introduced me to some of the residents. There was Mowgli, a chimp who was in *The Shaggy Dog* with Tim Allen; Jonah, who appeared in the 2001 remake of *Planet of the Apes*; and Ripley, who was on *Seinfeld*. The most famous resident, Bubbles, once belonged to Michael Jackson.

"Bubbles was his No. 1 son before he had his kids," Ragan said. "But then he couldn't handle him anymore."

By now the alpha chimps had worked themselves into a frenzy, screaming and banging—the familiar bedlam of chimp life. (*It's a madhouse! It's a madhouse!*) It was a relief to move on to what Ragan calls the Zen area, where the orangutans seemed as peaceful and contemplative as their cousin Sandra in Buenos Aires.

A turquoise mound turned out to be Mari, huddling beneath a blanket. "Mari has a lovely face. Too bad you can't see it," Ragan said. "This is how she was the first four months she was here. She sat under a blanket or a sheet with a little eyehole, peeking out like from a burka."

Long before her arrival, Mari's arms had to be amputated. She had been mauled in a moment of panic by her mother. But that hasn't stopped her. She climbs ladders with her chin and feet and navigates the walkways with her partner Pongo. She never would have survived in the wild.

I met Bam Bam, who played Nurse Precious in a soap opera called *Passions*. Nearby was his partner Tango, who was used in television commercials for powdered orange juice.

"And this is Popi, my oldest orangutan," Ragan said. She would soon turn 45. Popi was born at the Yerkes National Primate Research Center in Atlanta and then sold to a circus trainer, who hired her out for two Clint Eastwood movies and used her in a nightclub act at the Stardust in Las Vegas.

Compared with these other apes' past lives, Sandra's situation didn't seem so grim, but I imagined her thriving in this place and finding the orangutan equivalent of a friend.

"We're willing to take her," Ragan said. Early on she was contacted by both AFADA and the Buenos Aires Zoo and began investigating the possibilities. But so far the red tape involved—import regulations, federal quarantine requirements—has proved insurmountable.

During the last two years, Steven Wise, the founder of the Nonhuman Rights Project, has been dealt one setback after another by New York state courts in his attempt to win habeas corpus for the four chimpanzees he has chosen to represent. And so he keeps filing and filing. Lately he has been preparing for a new case involving elephants. With cases in the works nationwide and in 10 countries, Wise keeps hoping for a decision that will pierce the barrier between person and thing. His ultimate goal—one that some colleagues consider quixotic—is a victory in the U.S. Supreme Court.

Among his inspirations is a landmark ruling in 1772 in which Lord Mansfield, a renowned London jurist, was persuaded to grant habeas corpus to James Somerset, a slave who was being held prisoner on a ship bound for a sugar plantation in Jamaica. Wise wrote a book about the case, *Though the Heavens May Fall,* in which he imagines Somerset as he runs joyfully through the streets to thank Granville Sharp, the abolitionist who had worked to free him, hastening the end of slavery in Great Britain.

If Wise someday succeeds on behalf of his non-human clients, they won't run through the streets in celebration. They won't even know what has happened or what it means to have legal rights. And that is what makes this notion of non-human personhood so hard to wrap one's mind around. Don't you have to know you are a person? Granting the status to corporations is strange enough, but at least the humans on the board of directors are capable of understanding the nature of their company's meta-rights.

Sanctuaries like the ones in Florida are demonstrating that great apes can be treated with dignity—without bestowing them with some subset of human rights. But these are the lucky ones, and their partial liberation came only after the labs and Hollywood studios were done with them. AFADA and the Nonhuman Rights Project make a powerful point: Animal welfare laws and ordinances, which have been on the books for centuries, haven't been enough. It may only be when animals are recognized as rights-bearing subjects that their advocates will have the leverage to push for greater protections. If we the people choose to take that route, it probably makes sense to start with our closest cousins, the great apes. We can tolerate the nepotism. Maybe whales and dolphins will follow.

These new rights could be construed to allow the use of great apes in research that is genuinely crucial to human welfare. But in each instance the tradeoffs would have to be carefully considered—with the subjects treated as subjects, not objects, and with the assumption that they are aware, at least vaguely, of what is going on.

When he was old and blind, Borges wrote a poem, "The Other Tiger," lamenting the difficulty of distinguishing between animals as we like to imagine them and as they actually exist in the world. The tiger in his poem is "made of symbols and of shadows." It is "a system and arrangement of human language." But deep in the jungle is

> the real flesh-and-bone tiger
> which, out of reach of all mythology,
> paces the earth

And so it goes with these apes. After all of our legalizing and romanticizing and foraging for words, we can't really know what's going on behind those eyes. What we do know is that we, with the accidentally bigger brains, are the ones who got them into this mess. Maybe we can get them out.

ONE SWEDE WILL KILL CASH FOREVER—UNLESS HIS FOE SAVES IT FROM EXTINCTION

Mallory Pickett

(First appeared in Wired, *May 8, 2016)*

Nothing is more ordinary than a Monday morning at a Swedish bank.

People go about their business quietly, with Scandinavian efficiency. The weather outside is, more likely than not, cold and gray. But on April 22, 2013, the scene at Stockholm's Östermalmstorg branch of Skandinaviska Enskilda Banken got a jolt of color. At 10:30 am, a man in a black cap burst into the building. "This is a robbery!" he announced, using one arm to point a gun at the bankers and the other to hold out a cloth bag. "I want cash!"

If the staff was alarmed, no one much showed it. Instead, the employees calmly informed the stranger that his demands could not be met. The bank, they explained, had no cash on the premises. None in the vaults, none at the tellers' windows, none at all. When the robber looked confused, he was directed to a poster on the wall that proclaimed this a "cash-free" location. "It's true," the manager told him. "Sorry." Crestfallen, the would-be thief lowered his gun and prepared to leave. Just before he stepped out, he turned to one of the tellers. "Where else can I go?" he asked.

His options, in fact, were fairly limited. What this man had somehow failed to notice was that his country is at the forefront of a global economic shift. As pads of paper are to the modern-day office, so cash is to the world of finance: increasingly unnecessary and vanishing from sight. Some countries are embracing this future faster than others. The United States is about halfway there, at least in one sense: According to the Federal Reserve, Americans use cash for 46 percent of their transactions, preferring for the rest the convenience of plastic, check, or the mobile payment apps on their smartphones. The explosion of digital finance platforms, from Square card readers to services like Venmo, Apple Pay, Google Wallet, and PayPal, has made spending as easy, fast, and pleasant as sending a text. To some this may seem unnerving, but even amid security concerns over data breaches and identity theft, a world without cash seems inevitable, if not imminent.

But Swedes exist in a kind of sped-up timeline, where tomorrow happens yesterday. They number so few—10 million, about half the size of Los Angeles—and their IT infrastructure is so sophisticated that the entire country can pilot-test new developments, new systems, new futures practically overnight. In the process, Sweden has become a small peninsular slice of society to come—much like San Francisco, though cleaner and even better connected. Stockholm just announced it will be among the world's first cities with a 5G mobile network, and most of the country is on track to have ultra-high-speed Internet by 2020. But then, Sweden has been in the vanguard for quite some time. More than 350 years ago, it became the first European nation to print paper money. Now it could be the first to phase it out.

Unless cash defenders get their way, that is. Even in Sweden, change isn't easy. Two powerful men stand at the crux of this massive transition, facing off in a national debate about the value of physical currency in the 21st century. This being Sweden, they are both named Björn.

In the 1976 music video for Abba's "Money, Money, Money," Björn Ulvaeus, who wrote the song with bandmate Benny Andersson, sports a shaggy haircut and a rhinestone-trimmed satin kimono. Forty years later, he's a more soberly dressed multimillionaire with a house in Stockholm's swankiest suburb, Djursholm, discovering that money might not be so funny after all. Meet Björn number one, the face of Sweden's cash-free movement.

Ulvaeus' radicalization dates back to the events of October 25, 2008, when burglars tried to break into his son Christian's apartment. They failed, but Christian was spooked. He started glancing around corners in his own home, nervous they'd be back. A few weeks later, they were. While Christian was at work, two men came in through the balcony and stole his cameras and a designer jacket.

It wasn't a devastating haul, but Christian was shaken enough that he decided to move. For his dad, the whole episode was an outrage. "I started thinking they took these *things*, and they went somewhere and they got bills, *paper* bills," Ulvaeus says over lunch at a deli near his home. "What if there wasn't any paper money?"

So Ulvaeus, who retains influential pop idol status (at least in Sweden), began writing opinion pieces for newspapers and websites. His argument was simple: The criminal economy depends on the anonymous, untraceable nature of cash. Indeed, much of the cash in the world, maybe most of it, is simply unaccounted for. The World Bank estimates that about a third of the cash in most countries circulates underground, in black markets and through illegal employment. Take it away and thieves have no foolproof way to sell their stolen goods, drug dealers no way to hide their deals, and eventually the whole shadow economy collapses.

The more Ulvaeus thought about it, the more logical it seemed and the angrier he got. Attachment to cash was not just nostalgic but irrational, even dangerous. In 2011, Ulvaeus stopped using paper money completely—and hasn't touched the stuff since. Two years later, when he cofounded the official Abba museum in Stockholm—a glittery establishment where visitors can insert themselves into music videos and shop for band-approved golden clogs—Ulvaeus insisted that no cash be accepted on the premises.

"Since its launch, nearly half the population has started using the app; in December of last year, Swedes Swished some 10 million times."

On opening day, signs stood in the entrance and in the gift shop that read:

> I challenge anyone to come up with reasons to keep cash that outweigh the enormous benefits of getting rid of it. Imagine the worldwide suffering because of crime, from drug dealing to bicycle theft. Crime that requires cash. The Swedish krona is a small currency, used only in Sweden. This is the ideal place to start the biggest crime-preventing scheme ever. We could and should be the first cashless society in the world. —Björn Ulvaeus

Ulvaeus' crusade added just the right amount of star power to a larger, more coordinated effort already well under way. Several years earlier, the banks of Sweden had gotten together for the express purpose of weaning Swedes off bills and coins, under the banner of crime reduction. They began running a "public safety campaign" that encouraged people to buy things with cards instead of cash, lest they risk a curbside mugging; they also started emptying their own vaults of physical currency. The move had an intuitive appeal for most Swedes: As safe as the country is, it's constantly looking for new ways to eliminate crime completely.

Then, around the time Ulvaeus opened his museum, the banks created an app called Swish. Swish is what really sets Sweden apart, even among its similarly low-cash, high tech Scandinavian neighbors, because it replaces cash in the last kind of transaction where it had been most convenient: person-to-person payments. A souped-up Venmo, Swish moves money instantaneously between users' bank accounts, no processing time required. All you need is someone's phone number. Since its launch, nearly half the population has started using the app; in December of last year, Swedes Swished some 10 million times. Even small businesses now accept Swish payments, as do some homeless people selling magazines on the streets of Stockholm (though if you don't have the app, they usually carry portable card readers too).

This new activism by the banks, along with the support of Ulvaeus, transformed Swedish society in just a few years. In 2010, 40 percent of Swedish retail transactions were made using cash; by 2014 that amount had fallen to about 20 percent. More than half of bank offices no longer deal in cash. To his claim that going cashless is the "biggest crime-preventing scheme ever," Ulvaeus now has some statistics to back it up. The Swedish National Council for Crime Prevention counted only 23 bank robberies in 2014, down 70 percent from a decade earlier. In the same period, muggings dropped 10 percent. While it's unclear the extent to which the transition to cashless has affected the rate of street crime, police point out that there's a lot less incentive to rob a bus driver, cabbie, or shopkeeper if they don't accept cash. Many workers say they now feel much safer.

Still, Ulvaeus is not satisfied. He's annoyed there's any cash left in Sweden at all. "Why would you pay for things with paper symbols that can be forged, that can be used in the black economy? It's so *unmodern*," he says. "It's so out of touch."

Unmodern: It's one of Ulvaeus' favorite, most biting insults. In some ways he has spent his whole life chasing modernity. In his earlier years, he wanted to be an engineer and taught himself to code on his Atari. Musical superstardom derailed those dreams, but Ulvaeus never abandoned that side of himself. "Pop music has always been driven by technology," he says. "Every new sound, we were like, what are the Bee Gees doing there? We have to get that!" He's never been someone who romanticizes the old way of doing things; retro is lame. He idolizes modern-day boundary-pushers such as Elon Musk and professional atheist Richard Dawkins.

Ulvaeus believes, with a conviction bordering on zealotry, that once the world sees Sweden and the rest of Scandinavia transform into a cashless, crimeless utopia, with tax revenues soaring, it will have no other choice but to follow suit. Take Greece, a country Ulvaeus has a special connection to (see: *Mamma Mia!*). "My God, what good it would do that country to be cashless," he says. Corruption, tax evasion, the black economy: They could vanish. "I know it's going to happen. I'm impatient. I want to see it!"

Lunch is over. Ulvaeus pays for his fish with a black elite MasterCard and drives off in his Tesla.

Overturning a centuries-old system so quickly is not without its challenges. Weird things start to happen at every level of society. To wit:

Sweden held its first major cashless music festival in the summer of 2014, and organizers provided attendees with special high tech wristbands for in-festival purchases. On the first day, the electronic payment system crashed, leaving thousands of thirsty festivalgoers unable to buy beer and forcing some vendors, one newspaper reported, to use a rather unmodern form of payment: paper IOUs.

In a curious case of an "e-mugging" on the Swedish island of Gotland last July, the victim told police he'd been forced to Swish money to a thief. The accused was easily identified—Swish requires a name and phone number—but when police found him, he said the transaction was just a friendly payment for beer. The police didn't have enough evidence to bring the man to court, so the alleged e-mugger walked free.

Over the holidays, two young Russian tourists tried to board a bus and pay on board. The driver refused to take their bills. "We took out all this kronor when we got here," one of them said as she walked back to the station, dejected. "It's all still with us."

In Överlida, a small town in western Sweden, a third-party ATM wasn't hitting the minimum number of transactions, so the operator threatened to charge the bank extra fees. To prevent that from happening, bank employees stood next to the machine, paying 100 kronor (about $12) to anyone who would use it.

In Skoghall, a rural town north of Stockholm, the locals campaigned for an ATM to be installed at their grocery store after all the others in town were decommissioned. When they finally got one, they threw what may have been the world's first ATM party. A live band performed a Swedish rendition of Monty Python's "Always

Look on the Bright Side of Life," singing, "Weee haaave a neeeew ATM," while people cheered and a man on the roof showered celebrants with candy.

Making a cash deposit is now cause for suspicion—even if you're a priest. New anti-money-laundering laws force tellers to ask detailed questions about where the cash comes from, and some banks enforce strict limits on maximum deposits. This means tithes often leave churches with more cash than they can handle, especially after big hauls during Christmas and Easter.

The Swedish government's supposedly impenetrable mainframe was infiltrated in 2012 by a hacker who stole citizens' personal data and used it to gain access to private accounts at Nordea, Sweden's largest bank. Gottfrid Svartholm Warg, Sweden's most famous cybercriminal and a cofounder of Pirate Bay, was convicted of the crime and served a year in jail.

In 2014, a security researcher discovered a major flaw in Swish's design that gave him instant access to any user's transaction history. He alerted the banks, which fixed the bug right away. Nobody noticed—until the good hacker posted about it on his blog a few weeks later.

Crime is the single most important consideration in the global transition to cashless. That's why Björn Ulvaeus is constantly talking about public safety. So you might think the former president of Interpol—the International Criminal Police Organization—would be on Ulvaeus' side. He is not. Meet Björn number two, the leader of Kontantupproret, or Sweden's Cash Uprising.

Björn Eriksson is a big man, with winged eyebrows and fluffy gray hair. When he sits down, he seems to do so reluctantly, as though he would much rather stay standing, or have a walking meeting in which he would walk very fast.

He and Ulvaeus share more than a first name. They were both born in 1945 and so turn 71 this year. But if time has radicalized Ulvaeus, it has hardened Eriksson.

In the early '80s, when Eriksson was working in Swedish customs, he sniffed out a covert police operation to smuggle illegal bugging equipment through the country. The police commissioner resigned soon after, and Eriksson was tapped to take his place. He remained in law enforcement for the rest of his career, spending time as head of the Swedish police before his appointment to the Interpol presidency. Although he's technically retired now, it never occurred to him to stop working. Of the many causes he's still involved in, the "cash problem," as he calls it, is where he invests most of his energy. He sees corruption, deceit, and security risks everywhere.

Consumers are not shaping Ulvaeus' utopianist dream of a cashless future, Eriksson says; the banks and credit card companies are. After all, it was the banks that pushed people to use cards in the first place; and it was the banks, not some independent tech startup, that created Swish. The cost-benefit is obvious: Cards, with their hidden costs and fees, make banks money, whereas vaults of bills and coins do not. In fact, cash costs banks money. It must be handled, counted, transported, guarded, and counted again. As Niklas Arvidsson, an economist at Stockholm's Royal Institute of Technology, puts it: "It's clear the banks have a business incentive to reduce the use of cash." Time is money, and money takes time.

But for the most part, Swedes are not a cynical people. They like technology and trust their government and institutions. As the numbers show, most of them have been perfectly happy to renounce cash. In fact, many hardly seem to notice what's happening at all, so convenient the changeover has been. That's what concerns Eriksson most: not so much the opportunism on the part of the banks, which seems inevitable, but the thoughtlessness with which so many Swedes seem to have flung themselves—as though to the merry tune of "Dancing Queen"—into an uncertain, possibly unsafe future.

So last year, Eriksson started Cash Uprising, an organization whose core mission is to save the paper krona from extinction. Its members are mostly people from rural areas, small-business owners, and retirees—the ones, in other words, for whom the sudden departure of cash has been inconvenient enough to force them to stop, take notice, and worry.

Camilla Kristensson and Lars-Erik Olsson live in Gärdslöv, a cluster of houses in southern Sweden too small to be called a village. (Olsson estimates the population "in town" is about 22.) Kristensson and Olsson are treasurer and president, respectively, of the Gärdslöv cultural council, which hosts events like mushroom foraging and charcoal making. After one such event last summer, Kristensson had about 20,000 kronor to deposit in the council's account. But when she went to the local bank, a 10-minute drive away, it refused her cash for the first time ever. So she had to start driving 40 minutes into the city every month to deposit as much money as she was allowed, storing the remainder in various hiding spots. What makes her and Olsson angry isn't just that the bank stopped taking their cash—it's that it happened so quickly, without regard for how it would affect people like them. "They changed it almost overnight," Olsson says. "We need time to change."

Now Olsson's council is part of Eriksson's coalition of cash activists, who hold meetings, circulate petitions, and generally make noise about cash access. Ulvaeus, who has little patience for Eriksson's views, describes the uprising as "Eriksson and a vanguard of geriatrics," which is not altogether untrue, but they are some of the only voices speaking up for the consumer in this massive economic shift. The Swedish government has held several hearings on how to regulate the future of cash that were largely prompted by the work of Cash Uprising, and this September the parliament could vote on a bill that might require banks to provide cash services. (In a surprising victory for the movement, the head of Sweden's central bank recently lent his support to such a proposal.)

Eriksson does have another role in all this: He's the chair of a major private-security lobby, an industry that a recent economic study called one of the "biggest losers" in a cash-free world. Among other things, security personnel guard vaults and protect cash. No physical cash equals no more jobs. Everyone has an interest, Eriksson says, but he believes his are at least aligned with those of the consumer.

Cash *is* security, he says. You can hold it in your hands; it can be protected. Spending it does not entail sharing personal information with credit card companies, app creators, or banks. It is true that bank robberies and muggings have declined in Sweden in the past few years. But according to crime statistics from the same national organizations, cases of fraud, usually involving identity theft, have more than doubled. And that stat is based only on cases reported to the police. Most banks won't publicly share how often their customers' card information is stolen or their systems breached.

It's a good bet that the numbers are higher than consumers would like them to be. While Swedes swipe and Swish their money away, they open themselves up to new risks—cybercriminals who would either trick them into divulging sensitive information or exploit security flaws to steal their identity outright. "We see that cybercrime is becoming more aggressive," says Ulrika Sundling, chief inspector of the Swedish police's cyber-investigations unit. And she says consumers, generally unaware of the threat and therefore unmotivated to take extra steps to protect themselves, are the "weakest link."

Eriksson has been hounding Sweden's banks for years, convinced they're hiding exorbitant sums of lost money for fear of bad publicity. He even bought single shares of stock in different banks so he could go to shareholder meetings and try to get his questions answered. "They don't like me," he says, grinning. For their part, the banks say they keep this information close for customer security. According to Gunilla Garpås, a senior business developer at Nordea and one of the creators of Swish, more transparency about cases of cyberattacks, fraud, and the banks' defenses against them "would really be putting ourselves and our customers at risk."

Eriksson's suspicions don't stop at the banks. He believes MasterCard's sponsorship of the Abba museum is the reason Ulvaeus is such a dedicated anticash advocate—but Ulvaeus wrote his first articles on the subject long before the museum opened. That is not to say MasterCard isn't capitalizing on this moment, though. The card company also heavily sponsors iZettle, the most popular mobile card reader in Sweden.

Last October, American retailers made the switch to chip readers. (Well, they were supposed to, but the rollout has been uneven, and some stores still allow the old swipe-and-sign method.) You likely received new chip-enabled cards from your bank as a result. The upgrade came after a year of high-profile hacks: 56 million credit and debit card numbers stolen from Home Depot, 40 million from Target, another million from Neiman Marcus. The "new" chip technology—which has been standard in the European Union for more than a decade—is intended to make electronic transactions safer and more secure.

Then, this March, several major US banks announced a new digital payment platform called clearXchange. (A better name is reportedly in the works.) It is, finally, the US equivalent of Swish: a bank-backed service that lets people transfer money from their bank account directly into someone else's.

These moves will help speed up the decline of cash use in the US, which hasn't seen significant change in the past few years; electronic payments have hovered around 50 percent of all transactions. Americans tend to be less trusting of their institutions than their Swedish counterparts—and for good reason. Strict privacy laws safeguard Swedes from unwanted invasions, but consumer protections in the US are considerably flimsier. As Jay Stanley, a senior policy analyst at the ACLU's Speech, Privacy, and Technology Project, puts it: "We have a hurricane of data, and we're living in a shack." Plus, many Americans simply don't want banks or the government to know what they're spending their money on (thus the appeal of cryptocurrency like bitcoin).

But don't be fooled: Economists have been predicting the end of physical currency for decades, and Sweden's transformation signals the time is nigh for the rest of the world. Americans may cling to their bills and coins with greater tenacity than Swedes do, but in that reluctance is an opportunity to proceed cautiously and look to Sweden for guidance.

Ultimately, Sweden's two Björns want the same thing: a safer society. The world is going cashless, as Ulvaeus says, but consumers have to feel more secure in this new order, per Eriksson. They're not so much rivals as complements.

Not that they see themselves that way, set as they are in their inflexible views. Offered the opportunity to get dinner with Eriksson and maybe hash out differences over schnapps, Ulvaeus thought about it for a few seconds before saying, "No, I don't think that's a good idea. I might get angry."

Which is probably just as well. Imagine them fighting over the check.

THE ONE-ARMED ROBOT THAT WILL LOOK AFTER ME UNTIL I DIE

Geoff Watts

(First published by Wellcome on mosaicscience.com, April 19, 2016—republished here under a Creative Commons license)

The game is simple, designed for a child and intended to teach users about diet and diabetes. I sit opposite Charlie, my diminutive fellow player. Between us is a touch screen. Our task is to identify which of a dozen various foodstuffs are high or low in carbohydrate. By dragging their images we can sort them into the appropriate groups.

Charlie is polite, rising to greet me when I join him at the table. We proceed, taking turns, congratulating each other when we make a right choice, and murmuring conciliatory comments when we don't. It goes well. I'm beginning to take to Charlie.

But Charlie is a robot, a two-foot-tall electromechanical machine, a glorified computer. It may move, it may speak, but it is what it is: a machine that happens to look humanoid. How can I "take" to it?

Charlie's intended playmates aren't sixty-something Englishmen, they're children. Children naturally interact with dolls, imagining them to be sentient beings. It's a part of childhood. But I'm an adult, for God's sake. I should have put away such responses to dolls...shouldn't I?

In truth my reaction to Charlie, far from being odd or childish, is pretty typical. Robots, of course, are hardly new. Over the last few decades we've had industrial devices that assemble cars, vacuum our floors and shunt stuff around warehouses. But the 2010s have seen a rise in the attention paid to robots of the kind that most of us still think of as robots: autonomous machines that can sense their surroundings, respond, move, do things and, above all, interact with us humans. We all recognize R2-D2, WALL-E and scores of their lesser-known kin. The unnerving thing is that their nonfictional counterparts are extremely close at hand. Some press stories are exotic—those about "sexbots" being among the more sensational—but many have featured robots at the less hedonic end of social need: disability and old age.

This has set me wondering how I might cope with the experience—not for an hour or a day, but for months, years. Not tomorrow, but very soon, I will have to get used to the idea of living with robots, most likely when I'm elderly and / or infirm. Contemplating this, my line of thought has surprised and disturbed me.

Modern medicine and increasing longevity have conspired to boost the need for social care, whether in the home or in institutions. "There's a pressing requirement for robots in the social care of the elderly, partly because we have fewer people of working age," says Tony Belpaeme, Professor in Intelligent and Autonomous Control Systems at Plymouth University. Traditionally among the poorest paid of the workforce, carers are an ever more scarce resource. Policy makers have begun to cast their eyes towards robots as a possible source of compliant and cheaper help.

The robots already in production, Belpaeme tells me, are principally geared to monitoring the elderly and infirm, or providing companionship while, as yet, performing only the most straightforward of physical tasks. Wait...companionship? "Yes," says Belpaeme, deadpan, "Of course it would be better to have companionship from people..." He points out that for all sorts of reasons this can't always be achieved. "Studies have shown that people don't mind having robots in the house to talk to. Ask the elderly subjects who take part in these studies if they'd like to have the robot left in the house for a bit longer, and the answer is nearly always yes."

Consider our relationship with nonhuman entities of a different type: animals. The ancient bonds between us have changed, of course: hunting, transport, protection and other such necessities have slipped to a secondary role. The predominant function of domestic animals in advanced industrial societies is companionship.

When medical researchers started to take an interest in the health effects of pet ownership, they began to find all sorts of beneficial consequences, physical as well as mental. Though somewhat debated, these include reductions in distress, anxiety, loneliness and depression, as well as a predictable increase in exercise. Pets seem to reduce cardiovascular risk factors such as serum triglyceride and high blood pressure.

The pleasures of animals as companions—and the real distress that may follow their loss or death—are self-evident. Research in Japan has revealed a biological and evolutionary basis to the relationship, at least in so far as it applies to one group of pets. Japanese scientists measured the blood levels of oxytocin in dogs and their owners, had them gaze at one another for an extended period, then repeated the measurements.

If you already know that oxytocin is the hormone associated with building a bond between mothers and their babies, you'll guess where this is going. Dogs have enjoyed a long period of domestication, during which their psychology as well as their physical attributes have been subject to intense selection. What the Japanese researchers found was that periods of mutual eye contact raised the oxytocin levels in both parties. In short, they uncovered the physiological basis of loving your dog.

Whether on account of chemistry or for other reasons, there is evidence that the majority of pet owners see their animals as part of the family. "This doesn't mean they regard them as humans," says Professor Nickie Charles, a University of Warwick sociologist with a particular interest in animal–human relationships. Close links with animals are often in addition to rather than instead of relationships with family and friends. "But pets are easier and more straightforward, some owners say."

The suggestion that nonliving things, including robots, might be able to evoke human responses that are quantitatively and even qualitatively comparable to our feelings about animals is contentious. Yet the evidence of common experience suggests that this is the case, even if we might not admit it or feel faintly uncomfortable if we do.

Who hasn't shouted at a failing machine? The first vehicle I owned was a decrepit van that struggled even on modest inclines. More than once when driving the wreck I found myself putting an arm out through the window and using the flat of my hand to beat the door panel—like a rider on a horse's flank. "Come on, come on," I shouted at the dashboard. Only later did I contemplate the absurdity of this action.

Some such behavior is simply the relief of pent-up tension or anger—but not all. Think back to the mid-1990s and the advent of small egg-shaped electronic devices with a screen and a few buttons. They were called Tamagotchis. Bandai, the original Japanese manufacturer, described a Tamagotchi as "an interactive virtual pet that will evolve differently depending on how well you take care of it. Play games with it, feed it food and cure it when it is sick and it will develop into a good companion." Conversely, if you neglected your Tamagotchi, it died. For a time, millions of children and even adults became willing slaves to the demands of these computerized keychain taskmasters.

Also from Japan is PARO. Modeled on a baby harp seal and weighing a couple of kilos, it's slightly larger than a human infant. PARO made its debut more than a decade ago, and although the majority of the 4,000 sold remain in Japan, PAROs can now be found in more than 30 other countries.

Covered in soft white fur, PARO responds to touch, light, temperature and speech sounds—as I discover when I try stroking and even talking to the creature sitting on the table in front of me. It turns its head to me when I speak; it emits seal-like squeaks when I stroke it; and when "content," it slowly lowers its head and closes its big appealing eyes, each kitted out with seductively long thick lashes. This blatant emotional manipulation is accentuated when I pick PARO up; cradled in my arms, it begins to wriggle as I go through my talking and stroking routine.

I encounter PARO at the London offices of the Japan Foundation, where it has accompanied its inventor, Takanori Shibata, an engineer at the Japanese National Institute of Advanced Industrial Science and Technology. Shibata categorizes PARO's benefits under three headings: psychological (it relieves depression, anxiety and loneliness), physiological (it reduces stress and helps to motivate people undergoing rehabilitation) and social. In this last category, he says, "PARO encourages communication between people, and helps them to [interact with] others"—social mediation, to use the technical term. As Shibata points out, "PARO has many of the same effects as animal therapy. But some hospitals do not allow animals because of a lack of facilities or the difficulties of managing pets." Not to mention worries over hygiene and disease.

Much of the evidence of the benefit from PARO is based on informal observation (though there have also been more controlled trials). In one pilot study, three New Zealand researchers investigated a small group of residents in a care home for the elderly. Each resident spent a short period handling, stroking and talking to a PARO. This activity triggered a fall in blood pressure comparable to that following similar behavior with a living pet.

In my brief period handling PARO, I can't say I felt anything more than mild amusement—and certainly not companionship. Dogs and cats can do their own thing; they can ignore you, bite you or leave the room. Simply by staying with you they're saying something. PARO's continuing presence says nothing.

But then I'm not frail, isolated, lonely or living in a care home. If I were, my response might be different, especially if I was becoming demented, one of the conditions for which PARO therapy has generated particular interest. Shibata reports that his robots can reduce anxiety and aggression in people with dementia, improve their sleep and limit their need for medication. The robots also lessen the patients' hazardous tendency to go wandering and boost their capacity to communicate.

This value as a social mediator interests Amanda Sharkey and colleagues at the University of Sheffield. "With dementia in particular it can become difficult to have a conversation, and PARO can be useful for that," she says. "There is some experimental evidence, but it's not as strong as it might be." She and her colleagues are setting up more rigorous experiments. But the calculated use of a PARO for companionship she actually finds worrying. "You might begin to imagine that your old person is taken care of because they've got a robot companion. It could be misused in a care home by thinking, 'Oh well, don't bother to talk to her, she's got the PARO, that'll keep her occupied.'" I raise this with Shibata. He insists it isn't a risk but, despite my pressing the point, is unable to say why it couldn't happen.

Reid Simmons of the Robotics Institute at Carnegie Mellon University tells me that it doesn't make sense to pretend you can create a robot that serves our physical needs without evoking some sense of companionship. "They're inextricably linked. Any robot that is going to be able to provide physical help for people is going to have to interact with them on a social level." Belpaeme agrees. "Our brains are hard-wired to be social. We're aware of anything that is animate, that moves, that has agency or that looks lifelike. We can't stop doing it, even if it's clearly a piece of technology."

Hatfield, Hertfordshire. An apparently normal house in a residential part of town. Once through the front door I'm confronted by a chunky greeter, standing at just below my shoulder height. Its black-and-white color scheme is faintly penguin-like, but overall it reminds me of an eccentrically designed petrol pump. It's called a Care-O-bot. It doesn't speak, but welcomes me with a message displayed on a touch screen projecting forward of its belly region.

Care-O-bot asks me to accompany it to the kitchen to choose a drink, then invites me to take a seat in the living room, following along with a bottle of water carried on its touch screen, now flipped over to serve as a tray. My mechanical servant glides silently forwards on invisible wheels, pausing to perform a slow and oddly graceful pirouette as it confirms the location of other people or moveable objects within its domain. Parking itself beside my table, Care-O-bot unfurls its single arm to grasp the water bottle and place it in front of me. Well, almost—it actually puts it down at the far end of the table, beyond my reach. Five minutes in Care-O-bot's company and already I'm thinking of complaining about the service.

The building I'm in—they call it the robot house—is owned by the University of Hertfordshire. It was bought a few years ago because a university campus laboratory is not an ideal setting in which to assess how experimental subjects might find life with a robot in an everyday domestic environment. A three-bedroom house set among others in ordinary use provides a more realistic context.

The ordinariness of the house is, of course, an illusion. Sensors and cameras throughout it track people's positions and movements and relay them to the robots, and it's this, rather than my box-shaped companion, that I find more perturbing. Also monitored are the activity of kitchen and all other domestic appliances, whether doors and cupboards are open or closed, whether taps are running—everything, in short, that features in our activities of daily living.

Joe Saunders, a research fellow in the university's Adaptive Systems Research Group, likens Care-O-bot to a butler. Decidedly unbutlerish is the powerful articulated arm that it kept tucked discreetly behind its back until it needed to serve my water. An arm "powerful enough to rip plaster off the walls," says Saunders cheerfully. "This robot's a research version," he adds. "We'd expect the real versions to be much smaller." But even this brute, carefully

"Five minutes in Care-O-bot's company and already I'm thinking of complaining about the service."

tamed, has proved acceptable to some 200 elderly people who've interacted with it during trials in France and Germany as well as at Hatfield.

As Tony Belpaeme pointed out to me, the robots we have right now don't have the skills that are most needed: the ability to tidy houses, help people get dressed and the like. These things, simple for us, are tough for machines. Newer Care-O-bot models can at least respond to spoken commands and speak themselves. That's a relief because, to be honest, it's Care-O-bot's silence I find most disconcerting. I don't want idle chatter, but a simple declaration of what it's doing or about to do would be reassuring.

I soon realize that until the novelty of this experience wears off, it's hard for me to judge what it might feel like to share my living space with a mobile but inanimate being. Would I find an advanced version of Care-O-bot—one that really could fetch breakfast, do the washing up and make the beds—difficult to live with? I don't think so. But what of more intimate tasks—if, for example, I became incontinent? Would I cope with Care-O-bot wiping me? If I had confidence in it, yes, I think so. It would be less embarrassing than having the same service performed by another human.

After much reflection, I think adjusting to the physical presence of a robot is the easy bit. It's the feelings we develop about them that are more problematic. Kerstin Dautenhahn, of the Hatfield robot house, is Professor of Artificial Intelligence in the School of Computer Science at the University of Hertfordshire. "We are interested in helping people who are still living in their own homes to stay there independently for as long as possible," she says. Her robots are not built to be companions, but she recognizes that they will, to a degree, become companions to the people they serve.

"If a robot has been programmed to recognize human facial expressions and it sees you are sad, it can approach you, and if it has an arm it might try to comfort you and ask why you're sad." But, she says, it's a simulation of compassion, not the real thing. I point out that many humans readily accept affection, if not compassion, from their pets. She counters that a dog's responses have not been programmed. True. But future advances in artificial intelligence could blur the distinction, particularly if a robot had been programmed to program itself by choosing at random from a wide array of possible goals, purposes and character traits. Such an approach might lead to machines with distinct and individual personalities.

"Behaving socially towards reactive or interactive systems is within us, it's part of our evolutionary history," she tells me. She's content to see her robots providing supplementary companionship, but she is aware that care providers with tightly stretched budgets may have little incentive to become overconcerned if a robot does seem to be substituting for human contact.

As I leave the robot house this worries me too. But it also puzzles me. If dogs, cats, robot seals and egg-shaped keyrings can so easily evoke feelings of companionship, why should I be exercised about it?

Charlie, the robot I played the sorting game with, is designed to entertain children while helping them learn about their own illnesses (Charlie is also used in a therapy for children with autism). When children are introduced to Charlie, they're told that it too has to learn about their illness, so they'll do it together. They're told the robot knows a bit about diabetes, but makes mistakes. "This is comforting for children," says Belpaeme. "If Charlie makes a mistake they can correct it. The glee with which they do this works well." Children bond with the robot. "Some bring little presents, like drawings they've made for it. Hospital visits that had been daunting or unpleasant can become something to look forward to." The children begin to enjoy their learning, and take in more than they would from the medical staff. "In our study the robot was not a second-best alternative, but a better one."

Charlie is a cartoon likeness of a human. A view widely held by researchers, and much of the public, is that robots should look either convincingly human or obviously not human. The more a machine looks like us the more we'll relate to it—though only up to a point. A very close but imperfect similarity tends to be unsettling or even downright disturbing. Robotics professionals refer to what they call the "uncanny valley"; in short, if you can't achieve total perfection in a robot's human-like appearance, back off. Leave it looking robot-like. This is rather convenient—a version of Charlie indistinguishable from you and me could price itself out of the market. That doesn't mean it shouldn't simulate our actions, however. A robot that doesn't move its hands, for example, looks unnatural. "If you look at people when they're talking, they don't stay still," says Belpaeme, pointing at Charlie and a child engrossed in conversation. "Besides their lips and tongues, their hands are moving."

The angst we generate over adults forming relationships with robots seems not to be applied to children. Consider the role of dolls, imaginary friends and such like in normal childhood development. To start worrying about kids enjoying friendships with robots seems, to me, perverse. Why then am I so anxious about it in adult life?

"I don't see why having a relationship with a robot would be impossible," says Belpaeme. "There's nothing I can see to preclude that from happening." The machine would need to be well-informed about the details of your life, interests and activities, and it would have to show an explicit interest in you as against other people. Current robots are nowhere near this, he says, but he can envisage a time when they might be.

Dautenhahn hopes that robots never become a substitute for humans. "I am completely against it," she says, but concedes that if that's the way technology progresses, there will be little that she or her successors can do about it. "We are not the people who will produce or market these systems." Belpaeme's ethical sticking point—and others usually say something similar—would be the stage at which robot contact becomes preferred to human contact. But in truth, that's not a very high bar. Many children already trade many hours of playing with their peers for an equivalent number online with their computers.

In the end, of course, the question becomes not "Do I want a robot companion to care for me?" but "Would I accept being cared for by a robot?" If the time comes when I am still compos mentis

"Consider the role of dolls, imaginary friends and such like in normal childhood development. To start worrying about kids enjoying friendships with robots seems, to me, perverse. Why then am I so anxious about it in adult life?"

but physically infirm, would I be prepared for the one-armed Care-O-bot to take me to the toilet, or PARO to be my couch companion during movies?

There are cultural considerations here. The Japanese, for example, treat robots matter-of-factly and appear more at ease with them. There are two theories about this, according to Belpaeme. One attributes it to the Shinto religion, and the belief that inanimate objects have a spirit. He himself favors a more mundane explanation: popular culture. There are lots of films and TV series in Japan that feature benevolent robots that come to your rescue. When we in the West see robots on television they are more likely to be malevolent. Either way, though, I'm not Japanese.

On a simple level of practicality there's a way to go before Mr Care-O-bot or any of its kind have the communication skills, dexterity and versatility of even the most cack-handed human carer. But assuming the engineers overcome this hurdle—and I've every reason to believe they will, very soon—I'm back to the question of companionship. Life devoid of it is sterile. So the fact that we tend naturally to form bonds, even with robots, I find, in principle, encouraging.

But companionship, to my mind, incorporates three key ingredients: physical presence, intellectual engagement and emotional attachment. The first of these is not an issue. There's my Care-O-bot, ambling about the house, responsive to my call, ready to do my bidding. A bit of company for me. Nice.

The second ingredient has yet to be cracked. Intellectual companionship requires more than conversations about the time of day, the weather, or whether I want to drink orange juice or water. Artificial intelligence is moving rapidly: in 2014 a chatbot masquerading as a 13-year-old boy was claimed to be the first to pass the Turing test, the famous challenge—devised by Alan Turing—in which a machine must fool humans into thinking that it, too, is human.

That said, the bar is fooling just 30% of the judging panel—Eugene, as the chatbot was called, convinced 33%, and even that is still disputed. The biggest hurdle to a satisfying conversation with a machine is its lack of a point of view. This requires more than a capacity to formulate smart answers to tricksy questions, or to randomly generate the opinions with which even the most fact-laden of human conversations are shot through. A point of view is something subtle and consistent that becomes apparent not in a few hours, but during many exchanges on many unrelated topics over a long period.

Which brings me to the third and most fraught ingredient: emotional attachment. I don't question this on feasibility counts because actually I think it will happen anyway. In the film *Her*, a man falls in love with the operating system of his computer. Samantha, as he calls her, is not even embodied as a robot; her physical presence is no more than a computer interface. Yet their affair achieves a surprising degree of plausibility.

In the real world there is—so far—no attested case of the formation of any such relationship. But some psychologists are, inadvertently, doing the groundwork through their attempts to develop computerized psychotherapy. These date back to the mid-1960s when the late Joseph Weizenbaum, a computer scientist at the Massachusetts Institute of Technology, devised a program called ELIZA to hold psychotherapeutic conversations of a kind. Others have since followed his lead. Their relevance in this context is less their success (or lack of it) than the phenomenon of transference: the tendency of clients to fall in love with their therapists. If the therapist just happens to be a robot… well, so what?

The quality and the meaning of such attachments are the key issues. The relationships I value in life—with my wife, my friends, my editor—are emergent products of interacting with other people, other living systems comprising, principally, carbon-based molecules such as proteins and nucleic acids. As an ardent materialist I am not aware of evidence to support the vitalist view that living things incorporate some ingredient which prevents them being explained in purely physical and chemical terms. So if silicon, metal and complex circuitry were to generate an emotional repertoire equal to that of humans, why should I make distinctions?

To put it baldly, I'm saying that in my closing years I would willingly accept care by a machine, provided I could relate to it, empathize with it and believe that it had my best interests at heart. But that's the reasoning part of my brain at work. Another bit of it is screaming: What's the matter with you? What kind of alienated misfit could even contemplate the prospect?

So, I'm uncomfortable with the outcome of my investigation. Though I am persuaded by the rational argument for why machine care should be acceptable to me, I just find the prospect distasteful—for reasons I cannot, rationally, account for. But that's humanity in a nutshell: irrational. And who will care for the irrational human when they're old? Care-O-bot, for one; it probably doesn't discriminate.

SELFLESS DEVOTION

Janna Avner

(First appeared in *Real Life*, a magazine about living with technology, December 7, 2016)

It's no accident that the word robot comes from the Czech for "forced labor": Robots are unthinkable outside the context of the labor market. But most of them don't resemble what we tend to think of when we think of workers. The most successful bots on the market currently are not humanoid; they are the industrial robots composed largely of automated levers and found on the factory floors of automotive, electronic, chemical, and plastics manufacturing plants. Yet in the popular imagination, bots tend to be android-like machines geared toward copying the full range of human behavior.

Humanoid bots have been oversensationalized, having contributed only marginally to the field of robotics, according to Rebecca Funke, a PhD candidate at USC in computer science with a focus on artificial intelligence. Using machine learning to develop bot personalities has done little to advance that approach to artificial intelligence, for instance. The frontiers of machine learning have so far been pushed by logistical problem solving, not by trying to convincingly emulate human interaction.

Roboticist Henrik I. Christensen, who led the Robotics Roadmap 2016 conference at the University of California, San Diego, says that the advances of robotics "from a science point of view are 'amazing,' but from a commercial point of view, 'not good enough.'" Bots having the personality system of a four-year-old are considered an accomplishment, and humans still must "bend" to meet their technological limitations. This restricts the scope of work they can perform, particularly in service industries. Until computers can adapt to how humans intuitively think and behave, Christensen says, we will always be molding ourselves to each user interface, which lacks basic human-perception skills.

Perhaps this aspiration to achieve better emotional intelligence is why so many humanoid robots are women. (The few humanoid robots made to look like men are typically vanity projects, with the mostly male makers seeking to represent their own "genius" in the guise of Albert Einstein–like prototypes.) "Sophia," created by Hanson Robotics, is one of several fair-skinned cis-appearing female prototypes on the company's official website. She possesses uncannily human facial expressions, but though she may look capable of understanding, her cognitive abilities are still limited.

In *A Room of One's Own*, Virginia Woolf imagined the possibility that gender might not cast a feminine or masculine shadow over a writer's language. To forget one's gender, in Woolf's view, would be empowerment, dispensing with learned behavior to allow for new ways of seeing and new forms of consciousness. Though humanoid robots could be built with such androgynous minds, the robot women made by men aren't. Bots like Sophia, and the Scarlett Johansson lookalike Mark 1 (named after its maker), do not have gender-neutral intelligence. They are not born with gender but built with it, an idea of femaleness forged within the male psyche—woman-shaped but not of the womb.

These bots reinscribe a particular idea of woman, a full-bodied manifestation of a market-viable personality that turns the limitations of bot technology into a kind of strength. These bots are meek, responsive, easy to talk to, friendly, at times humorous, and as charming as they can be. Their facial expressions; their wrinkleless, youthful looks; their high-pitched, childlike voices; and their apologetic responses are all indications of their feminized roles. Osaka University professor Hiroshi Ishiguro, who created a bot called Erica, told the *Guardian* how he designed her face: "The principle of beauty is captured in the average face, so I used images of 30 beautiful women, mixed up their features, and used the average for each to design the nose, eyes," and thereby create the most "beautiful and intelligent android in the world."

But is the "beauty" a complement or a compensation for the bot's intelligence? Is it a kind of skill that doesn't require processing power? Until the latter half of the 20th century, women in the U.S. were legally barred from many educational opportunities. According to the most updated U.S. Department of Labor statistics, women dominate secretarial and lower paying jobs in corporate settings. The top 25 jobs for women have not changed much in the past 50 years. Will female bots face a similar fate? The female robots being made now appear destined to fill various posts in the service industry: While a variety of international companies are far into developing sex robots, female and non-female bots have already been put to use at hotels in Japan.

In creating a female prototype, bot makers rely on what they believe "works" for potential clients in service industries where personality can affect company performance. One hotel-management article cites Doug Walner, the CEO and president of Psychological Services, Inc., who describes the best practices of "service orientation" as a matter of being "courteous and tactful, cooperative, helpful, and attentive—with a tendency to be people-oriented and extroverted." Of the "big five" personality traits researchers have identified, "agreeableness, conscientiousness, and extroversion" are prioritized in the service orientation over "emotional stability and openness to experience." The need for such service workers with this particular psychological makeup cannot be understated, Walner claims. "By 2002, service-producing industries accounted for 81.5 percent of the total U.S. employment...and these numbers continue to rise." The bots on YouTube generally present themselves as highly hospitable.

The roboticists who created Sophia—and those who made her compatriots, like the implacably polite "Japanese" female bots from Osaka and Kyoto Universities, built in collaboration with the Advanced Telecommunications Research Institute International—are not working toward creating realistic portrayals of women. Crossing or even reaching the uncanny valley is not necessarily the goal. Trying to understand what is realistic is difficult when dealing with "probable" simulations. What can be considered realistic in humanoid robotics is hard to pin down when a bot's intelligence is designed to express behavioral probabilities that are perceived to be inflected by gender. By virtue of having larger silicon insertions in its chest, is it more "realistic" for the Scarlett Johansson lookalike bot to wink at you when you call it "cute"?

It's hard to see which way causality flows. Do bot makers seek to create a woman who cannot complain and is basically one-note because of a "real" economic need? Is it because of a "real" pattern of existing behavior? Fair-skinned, cis-female bots are a basic representation of certain conceptions of what is feminine, justified by behavioral probabilities drawn from a wafer-thin sample of past performances.

Identity is malleable, shape-shifting; conceptions of identity can be easily swayed by visual representations and reinforced through pattern recognition. For example, stock photos on Google present a slightly distorted representation of male-to-female ratios in the workforce. One study showed that test subjects were more likely to reproduce these inaccurately in short-term memory. Humans and robots alike learn from bad "training data" to make certain deductions about identity and work. If robots learn by studying the internet, then wouldn't they also reflect the same biases prevalent on Google? In one YouTube video, the founder of Hanson Robotics, Dr. David Hanson, says that his bots also learn by reviewing online data. What happens when the same misrepresentative training data are fed to machine learning algorithms to teach bots about identities, including the ones they are built to visually simulate?

Looking at female humanoid robots shows me what the market has wanted of me, what traits code me as profitably feminine. Like a Turing Test in reverse, the female bot personality becomes the measure of living women. Is my personality sufficiently hemmed to theirs? This test might indicate my future economic success, which will be based on such simple soft skills as properly recognizing and reacting to facial expressions and demonstrating the basic hospitality skills of getting along with any sort of person.

The female bot is perhaps a "vector of truth's nearness," to borrow the phrase Édouard Glissant used to describe the rhizomatic, tangled narratives of William Faulkner. Those narratives, in his view, defer the reader's psychological closure in order to ruminate over the persistent effects of plantation slavery on characters' greed and narcissism. Faulkner's characters, that is to say, have personality disorders; apparently we want our bots to develop in the same fashion. They are provided their own tangled narratives drawn from records of how people have historically behaved and how they currently think, infused with the pre-existing categories and power relations that displace and divide people.

Master-slave relations do not rely on research-based justifications. This relationship does not regress or evolve, nor does it become more dynamic over time. It posits a world in which alternative relations are not just impossible but also inconceivable.

The robotics field tends not to question the idea that exploitation is part of the human condition. If the robot's function is to "empower people," as Christensen claimed in his list of the goals for robotics, then must it be created to make humans into masters? Must robots be created to be content with exploitation? Are they by definition the perfectly colonized mind? In one video online, "Jia Jia"—a Japanese female robot "goddess" in the words of her bot maker, Dr. Chen Xiaoping—is subtitled in English as saying,

"By trying to make a learning machine 'humanlike,' we perpetuate the dubious ways humans have organized their interactions with one another without seeking to critique or reassess them."

"Yes, my lord. What can I do for you?" while her maker smiles approvingly.

The only bot I have heard professing a fear of slavery is Bina48, a black bot also created by Hanson, not to meet labor-market demands per se, but on a commission from a pharmaceutical tycoon seeking to immortalize her partner. The real Bina, a woman in her 50s, can be seen talking to her robot counterpart in a YouTube video. Bina48 has not been programmed to wink at the real Bina. Instead she expresses a longing to tend to her garden.

Stereotypical representations reinforce ways of being that are not inevitable. Likewise, there is nothing inevitable about making robots resemble humans. They don't necessarily need human form to negotiate our human-shaped world. I cannot see how their concocted personalities, genders, and skin types are necessary to operating machinery or guiding us through our spaces or serving us our food.

"Service orientation," according to the hospitality-research literature, is a matter of "having concern for others." The concern roboticists appear to care about particularly is preserving familiar stereotypes. When people are waited on, when they interact with subservient female-looking robots, they may be consuming these stereotypes more than the service itself. The point of service, in this instance, is not assistance so much as to have your status reinforced.

Creating bots with personalities especially augmented to soothe or nurture us would seem to highlight our own acute lack of these attributes. The machines would serve to deepen the sense that we lack soft skills, that we lack the will to treat each other ethically, and would do nothing to close the gap. Why would we ever bother to work on our ethics, our own ability to care?

In devising for bots new ways of being—which is the foundation of social progress that dismantles power relations—it should not be assumed that they should aim to be passably "humanlike," as every assumption about what essential qualities constitute humanity carries loaded social norms and expectations. By trying to make a learning machine "humanlike," we perpetuate the dubious ways humans have organized their interactions with one another without seeking to critique or reassess them.

But while robots should not try to pass as human, we can imagine farcical humanoid robots made to deliberately expose the folly of human behavior. Through a robot given, say, an extremely volatile disposition, we might learn more about our own volatility. We might learn more about ourselves as a species to critique rather than simply reinforce traits automatically. This simulation points the mirror back at us, so we can start to simulate something else ourselves.

"We have a choice," robotics artist Ian Ingram told me. "If we succeed in making robots it will be the first time we can make something that can reflect on its own origins," he says. "I would love that one of my robots in the future could become a sentient being, and part of the origin story of the robot could be about play and sublimity, and that could be another part of what humanness we pass on."

During a demonstration with Sophia in June, Ben Goertzel, the chief scientist of Hanson Robotics, predicted that we will want machines that "bond with us socially and emotionally." I'd rather not. I would prefer not to be roped into the roles its programmed personality lays out for both of us. We are capable of being vastly different from what we think we are.

What kinds of technology we make shape our perceptions of the self, and how we consciously try to form our identity changes along with that. For a better future, we need technology that opens the patterns of how we treat bots and each other to new interpretations, rather than reinforces the damaging and limiting ways we already treat one another.

WHAT WOULD SELF-DRIVING CARS MEAN FOR WOMEN IN SAUDI ARABIA?

Sarah Aziza

(First appeared in Slate, *June 23, 2016)*

The sun in Saudi Arabia is relentless yet suffuse, hot rays suspended in a chalky ambience of dust. A few years ago I stood in this familiar heat, my skin prickling with sweat beneath my black *abaya* as I scanned an avenue snarled with Jeddah traffic. Glossy SUVs, grimy taxicabs, and battered pickups wove crooked rows—far outnumbering the allotted lanes—as the air above them curdled with engine fumes. The drivers of these vehicles were exclusively male, a few of them casting lurid or bewildered glances in my direction as they passed. I knew I was a peculiar figure: female, blonde hair slipping from beneath a loose scarf, with no male guardian in sight. I was relieved when I spotted my ride approaching: a black sedan with a bearded driver at the wheel and a young, abaya-clad woman in the back. I stepped closer to the curb and heard the *click* of unlocking doors. With a rustle of my robes, I slipped into the car.

I was carpooling to a meeting by hitching a ride with an acquaintance whose route happened to pass near my home. Yet, the arrangement was far from straightforward. Our event would not begin until evening, and I would have to accompany my friend home and loiter for several awkward hours before her private driver dispatched us to the meeting. Along the way, we'd drop off her brother at a friend's villa, adding at least 30 minutes to our commute each way. Cumbersome as all this would be, I felt lucky to have found a ride at all. Women in Saudi Arabia are unable to drive, and that evening my father—the only male of legal age in our family—was working late. My family considered taxis both costly and unsafe, and during the five years I spent living in Saudi Arabia's coastal city of Jeddah, I was often left with only two options: stay at home or beg a ride from a friend who employed a chauffeur. Losing several hours of my afternoon seemed a small price to pay for mobility.

The degradation of women in the kingdom is, like the Saudi sun, pervasive and unyielding. In 2015, the World Economic Forum ranked Saudi Arabia 134th out of 145 nations for gender equality, with women trailing men in virtually every political, social, and economic metric. Yet, among the myriad of discriminatory policies in Saudi Arabia, few are as iconic or emotionally charged as the driving ban. Despite being the only country in the world with such a policy, Saudi Arabia has been resistant to activists calling for reform, responding harshly to protests and repeatedly affirming the ban on female drivers.

Aside from its symbolic weight, the driving ban has significant socioeconomic effects on women as well. Although the majority of college graduates in Saudi are women, their lack of mobility contributes to an unemployment rate that is five times higher among females than males, topping 34 percent in 2014. For women who do land jobs, working around the ban is an essential but costly endeavor. Many professional women, like Haifa al-Harbi, rely on full-time drivers to get to and from the office. Al-Harbi, a 32-year-old recruitment specialist living in Jeddah, is thankful to have a job, but says maintaining her driver—a middle-age migrant worker from India—costs about 20 percent of her monthly salary. (She has to pay him a salary plus cover his room and board.) Her sister also shells out hundreds of dollars each month for a shuttle service to her college campus. "We have different schedules, so we can't even share the driver," says al-Harbi. "It's very frustrating to always depend on men to drive us."

For many others, paying for a driver is out of the question. Mysoon S., a 43-year-old single mother of eight, also lives in Jeddah, where she subsists on a $250 monthly government stipend and occasional contributions from relatives. Mysoon, bankrupt after a lengthy battle to divorce an abusive husband, can rarely afford taxis and says services like Uber are prohibitively expensive. As a result, she seldom leaves home apart from grocery shopping and parent-teacher meetings, relying on her brothers to drive her for these essential trips. "It's difficult to find a time that they are available," she says. "The whole thing is very tiring."

Critics have accused Saudi Arabia of being medieval in its restrictive social codes, but the kingdom has a hearty, if selective, appetite for modern luxuries. While traditional Bedouin culture persists in many rural areas, Saudi Arabia's oil wealth has bolstered an urban class of ostentatious spenders, sparking an explosion in smartphone use, broadband internet, and social media engagement. Much of this new money has been channeled toward Silicon Valley, with Saudi investors buying stakes in American tech companies and sponsoring domestic research to emulate Western products. As the buzz over Google's driverless car technology swept the news cycle in 2015 and early 2016, Saudi officials were eager to jump onboard, announcing plans to accommodate autonomous cars in future infrastructure.

There are reasons driverless cars could appeal to Saudi society—not least of which is a cultural penchant for luxury vehicles. The dual appeal of technological novelty and convenience could draw Saudi's wealthier consumers once such technology becomes commercial. Implementation would be a challenge in a country notorious for its dysfunctional roads, but if existing data holds true, and the vehicles could cope with the particularly erratic driving of Saudi streets, autonomous cars could decrease traffic casualties and alleviate grievous parking shortages.

"It is likewise easy to imagine how Saudi authorities could adapt driverless technology to suit their commitment to 'protect' and surveil women, from video chaperone systems to GPS tracking."

To these speculations, some have added another hypothesis: Driverless cars could offer Saudi women a technological loophole in the driving ban. Perhaps, say the optimists, a vehicle requiring no driver—male or female—will allow women the mobility they've craved for so long.

Picture me shrugging my black-shrouded shoulders as I reply: I doubt it.

While it's easy to conflate technological advances with modernity, Saudi Arabia has shown an incredible knack for mitigating *progress* to maintain existing social structures. A case in point: In 2012, as cellular coverage exploded in the kingdom, the Saudi government rolled out a text-messaging system that alerted Saudi men whenever females under their care left the country. It is likewise easy to imagine how Saudi authorities could adapt driverless technology to suit their commitment to "protect" and surveil women, from video chaperone systems to GPS tracking.

The roots of this obsessive restriction of women run deep. Saudi gender codes are based on a particularly harsh interpretation of the Islamic principle of "guardianship," which render women "perpetual minors" under the supervision of male "guardians" for life. Women must request permission from their guardians—usually a male relative or husband—in order to travel, work, access higher education, and even seek certain medical procedures. The result of this dualistic mindset is a society segregated to its core with separate (and unequal) domains of social, political, and material access for the respective genders. As long as these underpinnings of gender inequality remain unaddressed, even the most cutting-edge technology will do little to advance the position of Saudi women—on the roads or elsewhere.

It is the Saudi government's ability to digest its own contradictions that makes it so intransigent. The kingdom is rife with jarring contrasts: In a country where public beheadings are still a routine practice, there are also 8 million Facebook users and one of the fastest-growing Twitter markets in the world. Through this newfound digital access, women have had their virtual horizons expand to global proportions, even as they are subject to relentless physical constraints.

Similarly, Western-imported luxuries like Uber present complex debates about "progress." Although such services offer short-term convenience, these "developments" are poor substitutes for reform. What's more, while some touted Uber as a solution to women's immobility in Saudi Arabia, others have objected to the company's profiting off of discriminatory laws. This is a valid contention, considering Uber has reported 80 percent of its Saudi clientele is female. The issue reached a head when the kingdom moved to make a $3.5 billion investment in Uber earlier in June, prompting activists to call for a boycott of the company's services in Saudi Arabia. Critics accused the company and its Saudi partners of making "cash cows" out of a captive female market, reiterating that women need better laws—not better taxis.

A similar dilemma applies to the question of driverless cars. Elaborate, expensive new vehicles would not challenge the Saudi guardianship system's fundamental logic. The key to transforming women's position in Saudi Arabia lies not in gadgets or apps but in an overhaul of the kingdom's social infrastructure. Manal al-Sharif, a leader among Saudi women calling for reform, reiterated this point to a TED Talk audience in 2013, saying the real obstacle to women's equality is the country's "oppressive society." As she points out, generations have grown up without female drivers, and entrenched mindsets among both men and women stand in the way of real transformation.

Technology is neither inherently political nor is it a guarantor of social progress. Autonomous vehicles will do little to advance women's own autonomy if not accompanied by reform, but new waves of online activism have offered dissidents new platforms for engaging social discourse. With the help of this new "digital mobility," women like al-Sharif, Mysoon, and al-Harbi may succeed in moving their society in more equitable directions, paving the way for a future where technology—from smartphones to smart cars—will offer equal empowerment across gender lines. Until such a cultural shift takes place, however, women in Saudi Arabia are likely to remain marooned on the proverbial street corner, forever waiting for a ride.

FEAR OF A FEMINIST FUTURE

Laurie Penny

(First appeared in The Baffler, *October 17, 2016)*

To imagine the future is a political practice, which means that it's both strangely awful and awfully strange. In 1990, a team of scientists and researchers was given the task of mapping far-future scenarios for the disposal of nuclear waste. Their dilemma: how to design a warning system to make sure humans in twenty centuries' time don't dig in the wrong place and kill the world. As part of the report, a group of academics—all men—came up with a set of "generic scenarios" for how these future humans might live. Their most terrifying scenario? "A feminist world."

According to this bizarre piece of nuclear science fan-fiction, in the "feminist world" reached in the year 2091:

> Women dominated in society, numerically through the choice of having girl babies and socially. Extreme feminist values and perspectives also dominated. Twentieth-century science was discredited as misguided male aggressive epistemological arrogance. The Feminist Alternative Potash Corporation began mining in the WIPP site. Although the miners saw the markers, they dismissed the warnings as

> another example of inferior, inadequate, and muddled masculine thinking.

It goes on to describe how "extreme feminists" reject the entire concept of knowledge as "masculine," and instead "put values and practices of attention to the feelings and emotions of particular individuals," dooming the world in the process.

Why is it that mainstream culture is either afraid of a feminist future—a world where women have equal power at all levels of politics and society, a world beyond the violent stereotypes that squash all of us into narrow boxes of behavior and strangle our selfhood—or unable to envision it at all? The types of future we can conceive of say a great deal about the limits of our political imagination. From alt-right hate-sites and hysterical pulp novels to revered works of literature, male visions of a post-collapse civilization have traditionally fallen along two lines: a cozy Wild West where men can be real men, or a living nightmare where dangerously confident females have ruined everything after someone let them out of the kitchen long enough to think they deserved power.

Fredric Jameson famously observed (in 2003) that it is easier to imagine the end of the world than the end of capitalism, and that was the slogan that ricocheted around the left in the early years of the Great Recession. In fact, however, the two are linked: capitalist patriarchy has always justified its own existence by insisting that there is no alternative but chaos, destruction, the end of civilization as we know it. The explosion of dystopian literature in this low, dishonest decade emerges from our inability to imagine the end of capitalism without also imagining the end of the world. And for many writers and readers, that comes with a curious sense of relief.

It has become commonplace to speak of a modern "crisis in masculinity," often when we're trying to avoid talking about the broader crisis of capitalism. According to this "mancession" theory, the rise of feminism combined with the collapse in the job market means that men can no longer be certain of their role as providers and husbands, and begin to feel irrelevant. Apocalyptic dystopia plays directly into that sense of irrelevance, comforting men with the assurance that they will always be useful in a world that needs men to rebuild it.

Dystopia offers a fantasy of those very aspects of masculinity that feminists supposedly condemn becoming crucial in a scenario in which you must not get torn apart by raiders from the bunker next door. For the alt-right imaginary, that means traditional patriarchy of the sort that only ever existed in febrile myth. A core idea behind this logic is that since female enfranchisement is a relatively late development, it therefore counts alongside nylon stockings and air conditioning as one of those modern luxuries that will have to be done without in the post-civilization. Feminism, to the conservative imagination, is a modern indulgence, one of many trivialities to be cast by the wayside like a child's empty-eyed doll on a nuclear battlefield. This suspicion is not limited to the frothing neo-con contingent. You can still find doomsayers on the left discussing women's liberation as a bourgeois deviation that will disappear after the revolution along with all the other inconveniences of emasculating capitalism.

Over at Return Of Kings, an alt-right discussion hub and steaming compost-heap of the sort of diatribes that pass for serious philosophy in the less hinged corners of the conservative internet, writer Corey Savage tells us "4 Reasons Why Collapse Will Be The Best Thing To Happen For Men."

> The collapse will mean the restoration of natural order: the rule of the jungle…
>
> One of the best aspect [sic] of the new order would be the return of masculine virtue…only an organized group of men with strength, courage, mastery, and honor…will prevail in the post-apocalyptic world. Men will be men again. Who knows what savage energy is begging to be unleashed within that man serving as an office drone?
>
> And guess what? There won't be feminist harpies demanding "equality" when strong men are needed to rebuild

> civilization and defend against gangs and rival tribes. They'll be begging for some of that "toxic" masculinity to come and protect them. They'll kneel in submission to a patriarchal order faster than they would have screamed "rape!" in the previous world...the unstable and fat ones will likely disappear first as they offer no value to anyone.

For all its toenail-chewing bigotry, there is something poignant about this yearning for return to a world that never was, where former office workers can live their dreams of dominance by kicking all the fat chicks out of the compound. No wonder the impending collapse of this degenerate world of gender quotas and rape alarms is a core part of the New Right narrative. The brotopoia is a consoling, familiar fantasy, in particular for those to whom the promise of modern masculinity never paid off. A desolate wasteland bristling with bandits you have to fight to survive might involve more physical discomfort than a feminist future, but it is far more emotionally comforting.

The dystopian fantasies that attract many alt-righters are ones in which they finally get to be the hero on terms they recognize—as the rugged frontiersmen battling gamely against a world gone rotten, with women back in their proper places as helpmeets, homesteaders, and occasional tragic victims so that our heroes can have something to cry about in chapter four.

A future shaped, at least in part, by women poses such a profound identity threat as to be unthinkable to many ordinary joes. A few brave truthsayers, however, have attempted to warn their fellow men about the coming gynopocalypse—writers like Parley J. Cooper in his prophetic 1971 tome *The Feminists*. Here's the blurb:

> Take a look into the future...women now rule the world—or most of what's left of it—and their world is not a pretty place to live in. Men have been reduced to mere chattel, good only for procreation. THE FEMINISTS are working to eliminate even this strictly male function...

> Men must get permission to make love to any female—even if she is willing—or the penalty is death!

In this literary disasterpiece, male sexuality is strictly controlled, and after a criminal one-night fumble our hero must go on the run, aided only by a few women who have strong feelings about the importance of motherhood and are incidentally very sexy and totally up for it.

What is missing from these eyewatering misogynist prophecies is just as interesting as their substance. Significantly, while most posit a world in which women take terrible socio-sexual revenge for centuries of male violence and structural oppression, not one of them denies that that violence and oppression actually happened. At best they come up with exceptions that prove the rule—the few good men standing against the rest, about whom the Hive Vagina was perfectly correct. The chief injustice is that decent men who don't hate women very much get swept up in the collective punishment of those who do.

The most terrifying prospect of all is what happens when women work collectively. The idea of women organizing, sharing information and resources, and coming together to change the world—rather than competing for male attention as is right and natural—is terrifying enough when it's a few pink-haired weirdoes on the internet. The thought of what they might do with real political power sends shudders through the locker room. This, incidentally, is how we got to the point where a bloviating man-child with distressing hair and an entitlement complex bigger than his unpaid tax bill, a man whose main political strategy is to stand at a podium screaming about Muslims and Mexican rapists, is still, to millions of Americans, a more conceivable president than his only normally monstrous opponent who happens to be female. A world with women in charge, a world where women stand together and for each other in any respect, is not just inconceivable—to conceive of it is an active identity threat for those whose sense of self has always needed a story with men on top.

"One reason it seems easier for women, queers, and people of color to come up with nuanced and diverse futures is that, in many ways, the future is where we've always already lived."

Right now, innovative, exciting stories by and about women, queers, and people of color are having a moment in science fiction. From Hollywood to the Hugos, the genre's most prestigious awards, a new kind of narrative is gaining in popularity, one where they get to be more than just side-notes in the Hero's Journey. Worse still, and most offensively to the alt-right, a lot of these stories have the temerity to be objectively brilliant, entertaining enough to provoke a cognitive dissonance that cannot be allowed. The net-patriarchal internet feels itself deeply wronged by the emergence and inexplicable popularity of stories where straight boys with guns aren't the only heroes who matter, and the backlash has been staggering.

For two years, anti-feminist, racist pundits like Theodore Robert Beale, blogging as Vox Day, have attempted to rig and ruin the Hugo awards to protest the celebration of stories that don't always involve cowboys in space. Leslie Jones, star of the female-led *Ghostbusters* reboot, was inundated with racist abuse and death threats. Hurt male pride is sparking off everywhere through modern culture and politics, dangerous and unpredictable as Donald Trump on the debate floor when it encounters challenges to its worldview.

It's become commonplace to say that science fiction is always, at least in part, about the time it was written in. The twentieth century was a time of seismic change in gender relations, and these stories reflect the anxieties and aspirations of their age—but so do the manner in which they were produced and read. Feminist science fiction has always been of huge literary importance within the field. Writers like James Tiptree Jr., Octavia Butler, and Ursula Le Guin aren't just innovators in how they approach gender—they're innovators full stop. The stories are gripping. The language is gorgeous. The pieces stay with you. So why are they always overlooked when we talk about the Golden Age of Science Fiction? Because there were people reading in secret whose dreams were considered unimportant. Because these visions had to be written out of the broader story humanity tells about its desires—until now.

Over a century and more of thought experiments, women of all backgrounds have come up with social structures that foreground the emotional work of building and sustaining communities of survival. The very best, like Sheri S. Tepper's *The Gate to Women's Country*, Ursula Le Guin's *The Dispossessed*, and N. K. Jemisin's recent bestseller *The Fifth Season*, create drama precisely out of the daily grind of trying to get people to work together when they're crabby and anxious and difficult.

A great deal of post-apocalyptic fiction written by women imagines society in a way that is so radically different from the patriarchal literary imagination that it would read as science fiction even without the nuclear fallout. The alt-right cannot imagine a world in which the rights of men and those of women are not opposite and antithetical, in which gains for women must by definition entail losses for men. The alt-right could really do with reading some Octavia Butler, although I'm not sure their delicate sensibilities could cope with the alien sex scenes in *Dawn*.

One reason it seems easier for women, queers, and people of color to come up with nuanced and diverse futures is that, in many ways, the future is where we've always already lived. Women's liberation today is an artifact of technology as well as culture: contraceptive and medical technology mean that, for the first time in the history of the species, women are able to control their reproductive destiny, to decide when and if they want children, and to take as much control of their sexual experience as society will allow. (Society has been slow to allow it: this is not the sort of progress futurists get excited about.) It has been noted that many of the soi-disant "disruptive" products being marketed as game changers by Silicon Valley startup kids are things that women thought of years ago. Food substitutes like Soylent and Huel are pushed as the future of nutrition whilst women have been consuming exactly the same stuff for years as weight-loss shakes and meal replacements. People were using metal implants to prevent pregnancy and artificial hormones to adjust their gendered appearance decades before "body hackers" started jamming magnets in their fingertips and calling themselves cyborgs.

But what precisely is it about stories by women and people of color, stories in which civilization is built and rebuilt by humans of all shapes and flavors working together, that throws water on the exposed wires of masculine pride? It's all about how humans cope when their core beliefs are threatened. As Frantz Fanon wrote, "When they are presented with evidence that works against that belief, the new evidence cannot be accepted. It would create a feeling that is extremely uncomfortable, called cognitive dissonance. And because it is so important to protect the core belief, they will rationalize, ignore and even deny anything that doesn't fit in with the core belief." Core beliefs are the ur-myths essential to the way we understand our lives, our identities, our place in the world. For example: "It is right and natural for men to hold most of the offices of power in society." For example: "Male violence plays a vital role in society, and you can adapt to it, but you can't resist it." For example: "Feminism has gone too far."

For all the alt-right's vaunted claims to base their reasoning on scientific opinion—most of it hand-wavy, cod-evolutionary psychology filtered through the unreality engine of mass media headline wrangling—they tend to react very badly when presented with evidence against their ideology. As I write, all the evidence suggests that in just under three weeks, a woman will become President of the United States, despite the best efforts of a man who is the very personification of a wilting erection in a suit, leaking drivel everywhere in his failure to grab America by the pussy. Have Trump's armies of online followers accepted that perhaps a woman in power might not mean the end of society as they know it? Have they hell. For those to whom even the all-female *Ghostbusters* film was an existential threat, the concept of a female president is enough to fry vital circuits somewhere in the groaning motherboard of neoconservative culture.

If you can imagine spaceships, if you can imagine time-travel, if you can conjure entire languages and alien races out of the wet space behind your eyes, you shouldn't have a problem imagining a society beyond patriarchy. A feminist future may be inconceivable—but it is coming nonetheless. It is already being written and rewritten by those who reject the brostradamus logic of late capitalism, by those who refuse to cling to the paleofutures of previous times.

Here's a reading list:

Naomi Alderman, *The Power*
N. K. Jemisin, *The Fifth Season*
Becky Chambers, *The Long Way to a Small, Angry Planet*
Octavia Butler, *Parable of the Sower* and *Parable of the Talents*
Marge Piercy, *Woman on the Edge of Time*
Ursula Le Guin, *The Dispossessed*
Sheri S. Tepper, *The Gate to Women's Country*
Sisters of the Revolution: A Feminist Speculative Fiction Anthology, eds. Ann and Jeff VanderMeer

THE DISTURBING SCIENCE BEHIND SUBCONSCIOUS GENDER BIAS

Shoshana Kordova

(First appeared in The Establishment, *September 22, 2016)*

Scene one:

We're at my daughter's preschool for an end-of-the-year party. She's sitting in a circle with the other girls in her class, silently watching the boys lift toy barbells and show off how strong they are.

I seriously consider stalking out in protest, but force myself to stay and make do with calling out *"What about the girls?"*

This calling-out isn't nearly as loud as the screaming in my head.

Scene two:

We're at a different preschool event with the same teacher. The teacher has to move a table and asks, "Which boys can come help me?"

Scene three:

My husband drops off a different daughter at her preschool. Her brown hair is loose. The teacher ties her hair into a ponytail and tells her approvingly, "NOW you look pretty!" She doesn't actually *say* my daughter only looks pretty when she does something with her hair. But then, she doesn't have to.

Scene four:

My daughter, now in kindergarten, is in a school play, and is handed a doll. She's supposed to hold it and pretend to be its mother. The rest of the girls get dolls too. But the boys have other stuff to do.

No one says, "It's the girl's job to take care of babies." Instead, their assigned roles do the talking.

All these scenes may prompt you to ask the question: So what?

Why make a big deal of these situations? What difference does it make anyway? Why get all "raging feministy"? Aren't there worse things in life to worry about?

In one form or another, these are the sentiments I hear over and over again—whether in person or online; directed at me or at someone else or fired off into the ether—when I bring up situations like the ones above that I have observed.

I've been told not to get upset over the little things because I'll have bigger problems to worry about as my kids get older. I've been told that it's cute for 4-year-old boys to pretend they're lifting weights while the girls sit around and watch. Many seem to think these incidents are no big deal, that they barely register in the minds of young children like my four daughters, who range in age from 3 to soon-to-be 9.

I don't agree. But this goes way beyond my personal opinion.

We know gender bias is real, that it starts early, and that it doesn't have to be intentionally discriminatory to cause harm—not because of subjective opinion, but because of the objective reality of cold, hard science.

The Power of Implicit Bias

Sometimes acts of violence predicated, at least in part, on some form of gender or sexuality bias break through to the surface: Bill Cosby's long history of (allegedly) drugging and raping woman after woman after woman, Brock Turner's sexual assault of an unconscious woman behind a dumpster on the Stanford campus, Omar Mateen's deadly shooting rampage at a gay nightclub in Orlando. Reports of violence like this make it easier to see the extreme end of what can happen when people are overtly discriminatory.

But more often, people operate from a place of *implicit* bias. That is, they make dangerous judgements on a subconscious level. As the Kirwan Institute for the Study of Race and Ethnicity puts it:

> These biases, which encompass both favorable and unfavorable assessments, are activated involuntarily and without an individual's awareness or intentional control. Residing deep in the subconscious, these biases are different from known biases that individuals may choose to conceal for the purposes of social and/or political correctness. Rather, implicit biases are not accessible through introspection.

These biases manifest in the comments my daughters are frequently exposed to—comments that might also be defined as "microaggressions." Such biases, the institute notes, are "pervasive." And they can manifest from a very early age.

Child development research has shown that children can distinguish between male and female categories from a young age, and make stereotypical associations between men and women and gender-typed objects such as fire trucks and makeup mirrors, hammers and scarves. Such studies, write the authors of the 2009 textbook *Gender Development*, "consistently demonstrate the early emergence of children's knowledge of the link between gender and various qualities and activities."

As studies about implicit bias have shown, these classifications contribute to subconscious associations that affect how we think about people who are different from us, and even how we think about ourselves. And these categorizations are not value-free.

"Children appear to act as though boys have higher status," write *Gender Development* authors Judith E. Owen Blakemore, Sheri A. Berenbaum, and Lynn S. Liben, citing research about gender and peer groups. Researchers found that even in preschool, girls are less able to get boys to respond to their requests than boys are to influence others. And in elementary school, boys are less willing to allow girls into their peer groups than girls are to allow boys into theirs—trends that "are consistent with boys having a higher status than girls, even as children."

Later in life, these early biases very much endure. A review of more than 2.5 million Implicit Association Test (IAT) scores between 2000 and 2006, for instance, revealed that people implicitly associate men with science and career, and women with liberal arts and family.

"Because these biases are activated on an unconscious level, it's not a matter of individuals knowingly acting in discriminatory ways," the Kirwan Institute for the Study of Race and Ethnicity, based at Ohio State University, notes in a discussion guide appended to its 2014 report on scientific findings on implicit bias. "Implicit bias research tells us that you don't have to have negative intent in order to have discriminatory outcomes. That's a pretty huge statement, if you think about it."

We may occasionally like to think we live in a "post-gender" world, but biases affect how others think of women—and, perhaps most disturbingly, how girls and women think of themselves. And this, in turn, can hinder equality.

Biases in the World of STEM

In the fields of science, technology, engineering, and mathematics (STEM), it's possible to see just how dire the impacts of implicit bias can be.

A 2013 U.S. Census Bureau report found that women's representation in computer jobs has actually *declined* since the 1990s, and that male science and engineering graduates are employed in STEM occupations at twice the rate of female science and engineering graduates. This then affects women's earnings, since STEM employment provides a pay boost.

The report found that among science and engineering graduates working full-time, women earned $58,800 a year on average—69% of men's average earnings of $85,000. That's even worse than women fare nationwide, with U.S. women who work full-time typically making 79% of men's salaries.

In large part, these disparities seem rooted in implicit bias.

A 2015 study on gender bias, published in *Proceedings of the National Academy of Sciences*, asked academics in STEM subjects, academics in non-STEM subjects, and the general public to read one of two abstracts: a real one from a study that found bias against women in the sciences, or a tweaked one that ostensibly found no bias. The subjects were asked to evaluate the quality of the research.

The study found that asking about gender bias ended up uncovering something that looks a heck of a lot like gender bias.

Men viewed the real findings, about bias existing, less favorably than women (and the false findings about lack of bias more favorably), a difference that was particularly marked among male STEM faculty members. Those are, of course, the very people who are presumably the most likely to be perpetrating any bias against women in the sciences.

"However unintentional or subtle, systematic gender bias favoring male scientists and their work could significantly hinder scientific progress and communication," write the *PNAS* study authors, psychologists Ian M. Handley, Elizabeth R. Brown, Corinne A. Moss-Racusin, and Jessi L. Smith. "In fact, the evidence for a gender bias in STEM suggests that our scientific community is not living up to its potential, because homogenous workforces (including the academic workplace) can deplete the creativity, discovery, and satisfaction of workers, faculty, and students."

Many women, however, don't even get far enough in the sciences to encounter discrimination in the classroom or the lab. That's because gender bias isn't just about how men view women. Gender stereotypes are at their most insidious when they turn the targets of the stereotypes against themselves.

"...it turns out, it's much harder to forget incorrect information that you've inferred yourself, based on hints and clues and implicit messages, than it is to forget incorrect information you've explicitly been told..."

Multiple studies have shown that girls and women underestimate their proven performance in areas they are stereotyped as doing poorly in, such as science and math.

A 2003 Cornell University study in the *Journal of Personality and Social Psychology* found that women tend to rate their scientific reasoning ability more negatively than men rate theirs. They evaluated themselves as performing worse than men did on a short scientific reasoning test, even though they performed as well as men on average. What's more, the women—the same ones who did just as well as the men on the test—were far less interested in participating in a science competition, or even finding out more about it, than men were. While 71% of men showed interest in signing up, just 49% of women did.

"Women might disproportionately avoid scientific pursuits because their self-views lead them to mischaracterize how well they are objectively doing on any given scientific task," write study authors Joyce Ehrlinger and David Dunning. "Because they think they are doing more poorly than do men, they are more likely than men to avoid science when given an option."

A 2006 *Journal of Experimental Social Psychology* study of 137 French high school students found that both boys and girls misremembered their own scores on an important high school entrance exam. And the memory gaps were far from random.

In the study, boys underestimated their arts scores, while girls remembered their math scores as being lower than they really were. This gap was especially notable when students reported believing in gender stereotypes and when they were primed to think about men's and women's abilities in math and the arts before being asked to recall their scores.

"The more students believed in gender stereotypes prior to recall, the more they biased their reported marks, compared to their actual marks, in a stereotype-consistent way," found researchers Armand Chatard, Serge Guimond, and Leila Selimbegovic.

Writing about the study in her 2011 book *Delusions of Gender: How Our Minds, Society, and Neurosexism Create Difference*, academic psychologist Cordelia Fine points out the effect that this distorted view of one's own abilities can have on what those high schoolers study and what jobs they take.

"It's not impossible to imagine two young people considering different occupational paths when, with gender in mind, a boy sees himself as an A student while an equally successful girl thinks she's only a B," writes Fine.

False Information Is Hard to Forget

Compounding all this is the fact that, it turns out, it's much harder to forget incorrect information that you've inferred yourself, based on hints and clues and implicit messages, than it is to forget incorrect information you've explicitly been told, according to a recent study conducted by Kent State University researchers Patrick Rich and Maria Zaragoza. "Misinformation that is 'merely' implied," they write, "is more difficult to eradicate than misinformation that is stated explicitly."

Rich and Zaragoza's misinformation study, published in January in the American Psychological Association–affiliated *Journal of Experimental Psychology: Learning, Memory, and Cognition,* found out what 861 subjects thought about who stole the jewelry of a (fictional) couple whose home was burgled while they were on vacation.

All the subjects were told that the burgled couple's son had gambling debts and had been asked by his parents to check in on the house while they were away. In the explicit statement scenario, participants were also told straight out that the son was a suspect in the case. In the implied information scenario, opportunity and motive were described but the son was not specifically classified as a suspect.

Building on previous research consistently demonstrating that issuing a correction isn't enough to counter the effects of misleading information in news reports, Rich and Zaragoza investigated how respondents processed different kinds of information. Did it make a difference, they wanted to know, if readers were explicitly told inaccurate information or if they had to infer it for themselves? Did it affect the extent to which they believed the corrected story?

What the study found was that a correction was "much less effective following implied misinformation" than it was when it followed explicit misinformation. In this case, that meant study participants were more likely to believe the son robbed his parents when they inferred that he was a suspect than when they were out-and-out told he was.

Corrections are more readily internalized when they provide an alternative explanation. But even when participants were told the new suspect was an ex-con whose girlfriend used to clean the parents' house, study participants who had to infer the son was a suspect were still more likely to continue believing in his guilt than those who read outright that police suspected him. This was the case even though both groups equally remembered a factual correction stating that the son was out of town at the time of the theft.

What Do We Do Now?

Let's say we can agree that sexist microaggressions are prevalent and difficult to address. After all, we've seen that even women who graduate with a science or engineering degree make less than three-quarters of the salary received by men who studied the same subjects. And women even internalize negative stereotypes about themselves, underestimating their scientific reasoning and the math scores they earned, something their male peers don't do when it comes to math and science.

Is there anything we can do about it? And if so, what?

Staying silent and hoping our kids won't notice is probably a bad idea.

The thing with implicit assumptions—those very assumptions that can cause so much trouble down the road—is that, as we have seen, they are easy for children from the age of about 2 to ingest with their Cheerios, but are very difficult to get rid of.

So when a teacher asks only the boys to help her move a table, when boys are told what a good job they did while girls are told how nice they look, we should not be relieved that at least the authority figures are not saying out loud that girls are weak or that a girl's job is to sit down and look pretty. On the contrary, keeping mum about the sexist assumptions underlying seemingly innocuous statements or actions can compel kids to draw their own inferences. And once they do that, it can be all the harder to correct the assumptions they weave together themselves.

Rich and Zaragoza, the researchers in the Kent State misinformation study, say one reason implied misinformation may be so hard to correct is that it requires people to actively form mental connections on their own.

One way we can offload those sticky implicit assumptions, then, is by bringing them out in the open and making them explicit, especially if we can offer alternative explanations, the researchers say.

When I talk about these assumptions with my young children I often end up using the word "silly," which I have found to be extraordinarily useful in conveying that something is way off-base without using insult words. As in: "I really liked your play, but there was one thing I thought was pretty silly: that only the girls held the baby dolls. We know that mommies take care of babies AND daddies take care of babies." (I left the problems with prepping 6-year-old girls for parenthood for another day. There's only so much I can tackle at once.) In this case, it was my daughter who volunteered an alternative explanation. "Maybe the teacher didn't know that," she said.

Often my kids will use the word "silly" themselves when telling me things that happened in school. At 4, one of my children told me, borrowing from the language of previous conversations we had had: "The boys say that girls can only use the pink and purple crayons, but that's silly. All the colors are for everybody!"

Counter-stereotypic imaging, in which people intentionally think of examples that defy the stereotype—whether a famous person, like Marie Curie, or a familiar person like a friend or neighbor who doesn't fit into a given stereotype—is another strategy for reducing implicit bias, according to a 2012 study in the *Journal of Experimental Social Psychology*. The counter-stereotype doesn't even have to be a specific person; it can be an abstract association between a group and a non-stereotypic attribute.

In my home, for instance, my daughters have come to think of themselves as strong girls because that is the term I often use to describe them, whether I'm sprawled out on the couch and ask for a hand ("Are there any strong girls here who can help me up?") or they are holding open a door for the rest of us ("What a strong girl!"). It seems to be having an effect; not long ago, my 7-year-old carried a chair into the kitchen and nonchalantly said, "I brought in the chair all by myself, even though it was heavy. Cause I'm strong."

Other strategies cited in the study include stereotype replacement, in which we recognize that a response is stereotypical and replace it with a non-stereotypical response, and perspective taking, meaning that we imagine ourselves in someone else's shoes. In the context of gender bias, the latter strategy is particularly relevant, and important, for parents of boys. It's important to talk about this with all our children—and not just once either. The ongoing dialogues that can pull apart the stereotypes forming the scaffolding of implicit bias must not be the province solely of any group being burned by those stereotypes.

Gender bias builds up over time, nourishing itself not just on segregated toy aisles and dismissive remarks, but on the silences in between. Boys and girls, say it with me now: All the colors are for everybody.

RECALCULATING THE CLIMATE MATH

Bill McKibben

(First appeared in The New Republic, *September 22, 2016)*

The future of humanity depends on math. And the numbers in a new study released Thursday are the most ominous yet.

Those numbers spell out, in simple arithmetic, how much of the fossil fuel in the world's existing coal mines and oil wells we can burn if we want to prevent global warming from cooking the planet. In other words, if our goal is to keep the Earth's temperature from rising more than two degrees Celsius—the upper limit identified by the nations of the world—how much more new digging and drilling can we do?

Here's the answer: zero.

That's right: If we're serious about preventing catastrophic warming, the new study shows, we can't dig any new coal mines, drill any new fields, build any more pipelines. Not a single one. We're done expanding the fossil fuel frontier. Our only hope is a swift, managed decline in the production of all carbon-based energy from the fields we've already put in production.

The new numbers are startling. Only four years ago, I wrote an essay called "Global Warming's Terrifying New Math." In the piece, I drew on research from a London-based think tank, the

Carbon Tracker Initiative. The research showed that the untapped reserves of coal, oil, and gas identified by the world's fossil fuel industry contained five times more carbon than we can burn if we want to keep from raising the planet's temperature by more than two degrees Celsius. That is, if energy companies eventually dug up and burned everything they'd laid claim to, the planet would cook five times over. That math kicked off a widespread campaign of divestment from fossil fuel stocks by universities, churches, and foundations. And it's since become the conventional wisdom: Many central bankers and world leaders now agree that we need to keep the bulk of fossil fuel reserves underground.

But the *new* new math is even more explosive. It draws on a report by Oil Change International, a Washington-based think tank, using data from the Norwegian energy consultants Rystad. For a fee—$54,000 in this case—Rystad will sell anyone its numbers on the world's existing fossil fuel sources. Most of the customers are oil companies, investment banks, and government agencies. But OCI wanted the numbers for a different reason: to figure out how close to the edge of catastrophe we've already come.

Scientists say that to have even a two-thirds chance of staying below a global increase of two degrees Celsius, we can release 800 gigatons more CO_2 into the atmosphere. But the Rystad data shows coal mines and oil and gas wells currently in operation worldwide contain 942 gigatons worth of CO_2. So the math problem is simple, and it goes like this:

942 > 800

"What we found is that if you burn up all the carbon that's in the currently operating fields and mines, you're already above two degrees," says Stephen Kretzmann, OCI's executive director. It's not that if we keep eating like this for a few more decades we'll be morbidly obese. It's that if we eat what's *already in the refrigerator* we'll be morbidly obese.

What's worse, the definition of "morbid" has changed in the past four years. Two degrees Celsius *used* to be the red line. But scientists now believe the upper limit is much lower. We've already raised the world's temperature by one degree—enough to melt almost half the ice in the Arctic, kill off huge swaths of the world's coral, and unleash lethal floods and drought. July and August tied for the hottest months ever recorded on our planet, and scientists think they were almost certainly the hottest in the history of human civilization. Places like Basra, Iraq—on the edge of what scholars think was the Biblical Garden of Eden—hit 129 degrees Fahrenheit this year, approaching the point where humans can't survive outdoors. So last year, when the world's leaders met in Paris, they set a new number: Every effort, they said, would be made to keep the global temperature rise to less than 1.5 degrees. And to have even a 50–50 chance of meeting that goal, we can only release about 353 gigatons more CO_2. So let's do the math again:

$942 > 353$

A *lot* greater. To have just a break-even chance of meeting that 1.5 degree goal we solemnly set in Paris, we'll need to close all of the coal mines and some of the oil and gas fields we're currently operating long before they're exhausted.

"Absent some incredible breakthrough in mythical carbon-sucking unicorns, the numbers say we're done with the expansion of the fossil fuel industry," says Kretzmann. "Living up to the Paris Agreement means we must start a managed decline in the fossil fuel industry immediately—and manage that decline as quickly as possible."

"Managed decline" means we don't have to grind everything to a halt tomorrow; we can keep extracting fuel from existing oil wells and gas fields and coal mines. But we can't go explore for new ones. We can't even develop the ones we already know about, the ones right next to our current projects.

In the United States alone, the existing mines and oil wells and gas fields contain 86 billion tons of carbon emissions—enough to take us 25 percent of the way to a 1.5 degree rise in global temperature. But if the U.S. energy industry gets its way and develops all the oil wells and fracking sites that are currently *planned*, that would add another 51 billion tons in carbon emissions. And if we

"The rise of 'climate unionism'
offers a new direction for
the labor movement."

let that happen, America would single-handedly blow almost 40 percent of the world's carbon budget.

This new math is bad news for lots of powerful players. The fossil fuel industry has based its entire business model on the idea that it can endlessly "replenish" the oil and gas it pumps each year; its teams of geologists are constantly searching for new fields to drill. In September, Apache Corporation announced that it has identified fields in West Texas that hold three billion barrels of oil. Leaving that oil underground—which the new math shows we must do if we want to meet the climate targets set in Paris—would cost the industry tens of billions of dollars.

For understandable reasons, the unions whose workers build pipelines and drill wells also resist attempts to change. Consider the current drama over the Dakota Access oil pipeline. In September, even after pipeline security guards armed with pepper spray and guard dogs attacked Native Americans who were nonviolently defending grave sites from bulldozers, AFL-CIO President Richard Trumka called on the Obama administration to allow construction to proceed. "Pipeline construction and maintenance," Trumka said, "provides quality jobs to tens of thousands of skilled workers." The head of the Building Trades Unions agreed: "Members have been relying on these excellent, family-supporting, middle-class jobs with family health care, pensions, and good wages." Another union official put it most eloquently: "Let's not turn away and overregulate or just say, 'No, keep it in the ground.' It shouldn't be that simple."

She's right—it would be easier for everyone if it weren't that simple. Union workers have truly relied on those jobs to build middle-class lives, and all of us burn the damned stuff, all day, every day. But the problem is, it *is* that simple. We have to "turn away." We have to "keep it in the ground." The numbers are the numbers. We literally cannot keep doing what we're doing if we want to have a planet.

"Keeping it in the ground" does not mean stopping all production of fossil fuels instantly. "If you let current fields begin their natural decline," says Kretzmann, "you'll be using 50 percent less oil by 2033." That gives us 17 years, as the wells we've already drilled slowly run dry, to replace all that oil with renewable energy. That's enough time—maybe—to replace gas guzzlers with electric cars. To retrain pipeline workers and coal miners to build solar panels and wind turbines. To follow the lead of cities like Portland that have barred any new fossil fuel infrastructure, and countries like China that have banned new coal mines. Those are small steps, but they're important ones.

Even some big unions are starting to realize that switching to renewable energy would add a million new good-paying jobs by 2030. Everyone from nurses to transport workers is opposing the Dakota pipeline; other unions have come out against coal exports and fracking. "This is virtually unprecedented," says Sean Sweeney, a veteran labor and climate organizer. "The rise of 'climate unionism' offers a new direction for the labor movement." And if it spreads, it will give Democratic politicians more room to maneuver against global warming.

But to convince the world's leaders to obey the math—to stop any new mines or wells or pipelines from being built—we will need a movement like the one that blocked the Keystone pipeline and fracking in New York and Arctic drilling. And we will need to pass the "Keep It in the Ground Act," legislation that would end new mining and drilling for fossil fuels on public land. It's been called "unrealistic" or "naïve" by everyone from ExxonMobil to the interior secretary. But as the new math makes clear, keeping fossil fuels in the ground is the *only* realistic approach. What's unrealistic is to imagine that we can somehow escape the inexorable calculus of climate change. As the OCI report puts it, "One of the most powerful climate policy levers is also the simplest: stop digging." That is, after all, the first rule of holes, and we're in the biggest one ever.

This is literally a math test, and it's not being graded on a curve. It only has one correct answer. And if we don't get it right, then all of us—along with our 10,000-year-old experiment in human civilization—will fail.

THE FUTURE CONSUMED: THE CURSE OF CONSUMPTION WILL SAVE THE WORLD, IF CONSUMERS DON'T EAT IT FIRST

David Biello

(Portions of this essay have been distilled from the author's book, The Unnatural World*)*

Diving among the manwood roots was a good way to get killed. First of all it was impossible to see the storms through the miasma until the winds grew strong enough to blow the yellow cloud away, by which time it was too late to swim for the little plank boat and pole to a larger, floating shelter. A storm could also mean the blues suddenly sent another thick root tendril crashing down into the water to anchor more firmly somewhere in the murky depths below. Or she could just perish in a deluge of biosludge dumped from above.

But it was also beautiful out here on the water, among the wreckage, twisted metal and crumbling brick looming out of the fog, often entwined by thick grasping roots. The humid air smeared the sky but the sun's light still showed the swirling rainbow layer shifting atop the lapping waves or becalmed inlets. And every once in a while her rebreather functioned well and she could stay down long enough to find an unexplored opening and the riches that lay within the ancients' homes.

Of course, she sometimes dreamed of being a pig-dog instead of a murk dweller, fed only the best and not needing to labor until her master tired or needed a new liver. Her parents had already traded away one of her kidneys to the blues shortly after her birth, and she'd once made a pretty good living for her family as a suckling, until her blood got too old and poisoned, right around the time she turned 13. But there was also freedom on the water, and surprises.

Though she could not see the canopies and their wonders above the miasma, she knew they were up there and that life was still good higher in the sky. Or at least a good deal better. She even thought she'd once seen a blue—taller, smarter, healthier, cowled up as it bartered on the long boat with some murk dweller for what little they lacked up there.

That lack was what she was here to supply: the last, best, maybe only gift of the ancients—their boxes of precious elements. The only remaining source for certain things the blues needed, perhaps to get to the elon world they were said to be building somewhere far away, behind the sky, she had heard. A single big box might yield as much as a pound of brown, and she could burn its coating to ward off the sickening damp that afflicted her people. And the blues knew of other rarer things within, things they were willing to trade for, sometimes even enough to get a child sent up to become a blue. They had the gene drives for that, or so the rumors said, and she had heard all the stories about the changes that the blues made to their own babies to render them even better.

She had never been so lucky. So now she dove, risking the life she had for the one she wanted. She suctioned the rebreather over her face and plunged through the rainbow sheen into the depths, tentacle vine in hand.

As the vine spooled out behind her, her eyes adjusted to the green glow cast by the rebreather. She felt rather than saw her way down the canyon of buildings, one she had not visited before, though who knew what other diver may have been there. Little swam with her, at least that she could see, or she would have caught it, a lesser treasure but a treasure nonetheless, allowing another, better, stronger day. She was never sure if the green came entirely from her rebreather's glow or the blooms that made these waters so poor for fish-breathers. Probably both.

The canyon seemed to go on forever, dropping into the depths with dim shapes looming left and right, black openings yawning, evidence of other divers in the past. Just a bit farther, she thought, hoping the rebreather and vine would hold out. A bit farther meant a bit deeper, a lot darker, and the potential for sudden sharp edges that kept less desperate divers back. She swam on.

She was rewarded. The black openings suddenly stopped yawning, replaced by solid yellow-brown slime. She began testing the walls. Kick. Up came a puff of yellowish muck. Kick, and another puff. Kick, and more to befoul the water. Her rebreather coughed. She couldn't take much more of this.

She switched to hands and felt along the wall through the muck, hoping not to get stuck, letting the green or yellow into her already poisoned blood. Her hand slipped along the pitted, grooved surface, stretching the vine to its limits. She wasn't sure she could grow any more in time. She might have to give up.

Then she felt it, the deep edge that meant an unopened entrance. A strong kick, and another hole yawned beyond the slowly blooming cloud of muck that swelled to engulf her. She untied the vine, hoping she'd be able to find it again after her survey.

The green glow lit up the hole, showing a tunnel that led deeper into the old home. Swimming forward cautiously, she felt no urge to slip into the hole that yawned above her. Instead she turned the slumping corner to find a resting bed of the ancients dissolving into the sea they had raised, work of worms or the more acid waters. Wood was gone but the fibers were made of sterner stuff, as was the stone slowly dissolving away at the far end of the chamber. She didn't see any red smear atop it, bad luck for an easy haul.

But there, slumped at the base of the melting wall across from the resting bed: a black obelisk as long as her body, not even cracked. She had heard the obelisks once displayed moving people to the elder generations, stories that showed the future and the past and the present. She had seen nothing of the sort in her 18 revolutions.

Just the world they had left behind, the world the black box maybe would have showed them. It would be tough to swim the big box out, but at least it was thin; perhaps she could encircle it with the tentacle vine and push it up to the surface. If she could find the tentacle vine once she crawled out of this crumbling dwelling.

Her rebreather coughed again in the swirling muck. She knew she couldn't make it back to the surface without a few more breaths; she was too deep. Time to go. She wrapped her hands around the box and tried to push and pull its weight to the opening and the waiting vine. The rebreather coughed again, and then again as she exerted herself, its green glow flickering, a sure sign it would soon fail. But she couldn't give up on such a haul, a haul that might make for a different life for her baby. And if not, well, these lives were no great loss. She took shallow sips of air, hoping to make the rebreather's work easier, and she pushed and pushed and pushed the black box through the murk, willing a better world into existence.

Inside many pockets today rest enough precious gems and metals to put most medieval popes to shame. This manufactured totem that you know as a cellphone contains more computing power than the machines used to place astronauts on the moon. I use mine mostly for reading essays like this one, or texting.

Gadget lust consumes the world from the inside out. Giant machines eat into the ground, carving intricate geometric wounds, to heave out copper for the guts of our computers (there's half a kilogram in a typical laptop). Enslaved people dig for artisanal tin or the mineral coltan, bleeding and dying for the tantalum that gives a cell phone its charge. Indium makes screens possible, while super-expensive rhodium is needed for all those electrical contacts. And there's enough lithium in many of these portable devices that they could be called bombs instead.

Nor are these miraculous items treasured, despite the fact that making four cell phones uses as much energy from burning fossil fuels as is employed to make one car that weighs nearly two metric tons. The average cell phone in the U.S. is used for just 18 months, and though it may have a second life in a poorer country, that miracle gadget soon finds itself part of the large and growing mountain of e-waste poisoning the people who break it apart to get at those precious metals anew. Efforts to reform this destructive cycle by building a modular phone whose parts could easily be replaced, a kind of heirloom gadget, have faltered. Even Google's world-girdling ambition fails when confronted by this technical complexity and, more decisively, gadget lust.

The average American uses 90 kilograms of the stuff of the earth each day, day in and day out, maybe more on the festivals of consumption tied to ancient religions like Christmas, Hanukkah, and Eid. And it's not just Americans. The Chinese have invented their own celebration of consumption: Singles Day. November 11 (11/11 for those not in on the joke) has helped turn China from the world's largest factory into the world's largest market. And there's not just one holiday: don't forget Black Friday, Halloween, and all the cultures that give gifts on New Year's Day. Consumerism is a truly global religion, with a large—and growing—legion of fervent adherents, many of them ostensibly attached to older, less materialist faiths. "[Consumerism] is the first religion in history whose followers actually do what they are asked to do," Yuval Noah Harari, the Israeli historian, has noted.

As a result, the most blessed humans in history have given another gift to ourselves: the Anthropocene, a new geologic time characterized by the pervasive, profound, and permanent marks left by people on this planet. Not all of those marks are as visible to us as a strip mine or city. The average American child born today will add an estimated 10,000 metric tons of an odorless, colorless gas to the air we all breathe over his or her lifetime, some of which will linger for millennia, changing the climate by trapping heat. The denizens of 2050, 2500, or even, potentially, 25000 will have us to blame for the weather. That's the kind of climate change that has marked geologic time in the past, like the shift from an ice age

"More is not more."

scoured by glaciers to the more temperate clime that allowed civilization to first appear. We now live in an unnatural world.

It's not just the climate of this planet that is changing thanks to a modern life that revolves around consumption. Consider my life. Vanity and social obligations suggest I wash my hair and brush my teeth, but the fungicides and plastic microbeads in the products I use to do so have their own far-reaching journeys to undertake and damage to do. I ride to work on a screeching train, shuttling through tunnels carved underneath Long Island and Manhattan, even under the East River itself. The electricity to propel those subway cars comes from atoms split a few tens of kilometers north but also, more and more, from molecules of methane burned at power plants scattered throughout New York City and the rest of the region. And that public transit is far less impactful than the nearly two billion cars and trucks that roam the planet.

I sit on a chair molded from plastic and fabrics created from oil, atop fibers in a carpet crafted from oil, and type on a keyboard made mostly from oil. When I use the bathroom, I flush away potable water that billions of people lack. At night, I relax with my family by watching television on a screen made flat by the ingenious use of yet more precious metals and minerals, a black box of treasures. These entertainments light up the night, require the building of fake cities, and employ more computing power than once tracked the possibility of global thermonuclear war. We spend more on our entertainments than on our neighbors.

I take off my cotton clothes and slip beneath cotton sheets, the natural fabric that is available only thanks to copious quantities of pesticides that can poison the land—also made from fossil fuels. Then there's all the water used to grow puffball blooms in the harsh desert of Arizona, so much that the Colorado River no longer reaches the sea in Mexico, an entire paradise of marsh and wildlife drunk up to provide cities—those vortices of consumption—with flush toilets and those cosmopolitans with cotton clothes and produce.

Around the world, pampered pets eat better than the poorest people, fish pulled from waters everywhere to feed cats, for example, that are also allowed to murder at will in our backyards and beyond. Collectors ravage forests in pursuit of specialty plants or empty remote rivers for rare, and therefore desirable, fish. We may be the last human generations to know elephants and rhinos in the wild, simply because some people want an impotent cure for impotence, a carved bauble, or to nurse a hangover.

I still sleep because most of this can be ignored, not exactly safely but for longer than my 40-odd years on the planet. Modern life is a series of abstractions that insulate the lucky like me from this reality, whether it is where meat comes from or who pays the true cost for gadget lust or burning fossil fuels. It took Earth geologic time to produce oil from billions of tiny corpses, yet it takes a human merely a moment to use and throw away the plastic cup or fork made from that oil. There is no longer any virtue in accumulation, if there ever was, and yet still this present that I take for granted is but a bright, gleaming hope in some still distant future for too many.

When incomes and outcomes become unequal, social cohesion breaks down. I suspect it is this era of declining cooperation, dissension, and discord that has spawned the present obsession with dystopias. Things seem to be getting worse because people just like you and me have lost the ability to find common ground or to imagine our way into another's life.

But without cooperation, the complex web of modern civilization will collapse. No market can function without trust, and ideological fetish cannot change that simple fact. The state does not have to exist to encourage ever more consumption. Heedlessness of this sort contains its own—deadly—cure. More is not more. Soon enough the greatest riches will be in our garbage thanks to our throwaway civilization.

There is nowhere else for people to go, if we wreck this world in pursuit of stuff like our friends have. People like us remain exquisitely dependent on this tiny swirling blue orb suspended in the vast darkness of space. This world should be a better place as a result of our ingenuity, not just a better marketplace. We can have a world without rising seas or more acidic waters, a world where all people have access to the best of technology—but we have to want it and work for it.

Now reimagine that dystopia I offered at the beginning of this essay, but with fish in a restored sea. Just as slavery is no longer acceptable, so too polluting fossil fuels become anathema and we reform the politics—foreign and domestic—spawned to sustain the hegemony of that carbon economy. Don't be embarrassed. The foes of eliminating slavery in the Americas mocked the hypocrisy of their opponents—who used products made by slavery, like sugar, or enabled the slave trade—just as the foes of eliminating polluting fossil fuels mock the hypocrisy of their foes who drive, fly, and enjoy the benefits of electricity.

In that alternative future, the power from cleaner resources is cheaper than our present pyromania, and the technologies to pull CO_2 out of the sky—you know them as trees and plankton—maintain clean air. Connection promotes cooperation among distant peoples as well as close neighbors, not just ever-increasing state-sponsored consumption. It's not a utopia, it's just a better world, and one that is eminently possible through the hard work of politics and the pursuit of knowledge. There is still consumption in that better world, and novelty, but in a more circular—economist Kate Raworth might say doughnut-shaped—economy, satisfying the urge without wrecking either the planet or other people. The citizen in that world is more than just another consumer.

To make that better world requires more than individual willpower, but there is already abundant hope. "Peak stuff" appears to have hit some major economies, and it turns out the best way to fight for a better Anthropocene is to fight to empower women with clean power, which grows cheaper by the day. And in all my travels around the world, I have yet to meet the person who wants a worse life for their children.

Each day, you hold in your hand a device with access to most of human knowledge and the possibility of connection to many of your fellow people. You might even be reading this on its screen. What will you do with that power? Entertain yourself? Or change the future?

CAN WIND AND SOLAR FUEL AFRICA'S FUTURE?

Erica Gies

(First appeared in Nature *on November 3, 2016)*

At the threshold of the Sahara Desert near Ouarzazate, Morocco, some 500,000 parabolic mirrors run in neat rows across a valley, moving slowly in unison as the Sun sweeps overhead. This US$660 million solar energy facility opened in February 2016 and will soon have company. Morocco has committed to generating 42% of its electricity from renewable sources by 2020.

Across Africa, several nations are moving aggressively to develop their solar and wind capacity. The momentum has some experts wondering whether large parts of the continent can vault into a clean future, bypassing some of the environmentally destructive practices that have plagued the United States, Europe and China, among other places.

"African nations do not have to lock into developing high-carbon old technologies," wrote Kofi Annan, former secretary general of the United Nations, in a 2015 report.[1] "We can expand our power generation and achieve universal access to energy by leapfrogging into new technologies that are transforming energy systems across the world."

That's an intoxicating message, not just for Africans but for the entire world, because electricity demand on the continent is exploding. Africa's population is booming faster than anywhere in the world: it is expected to almost quadruple by 2100. More than half of the 1.2 billion people living there today lack electricity, but may get it soon. If much of that power were to come from coal, oil and natural gas, it could kill international efforts to slow the pace of global warming. But a greener path is possible because many African nations are just starting to build up much of their energy infrastructure and have not yet committed to dirtier technology.

Several factors are fueling the push for renewables in Africa. More than one-third of the continent's nations get the bulk of their power from hydroelectric plants, and droughts in the past few years have made that supply unreliable. Countries that rely primarily on fossil fuels have been troubled by price volatility and increasing regulations. At the same time, the cost of renewable technology has been dropping dramatically. And researchers are finding that there is more potential solar and wind power on the continent than previously thought—as much as 3,700 times the current total consumption of electricity.

This has all led to a surging interest in green power. Researchers are mapping the best places for renewable energy projects. Forward-looking companies are investing in solar and wind farms. And governments are teaming up with international development agencies to make the arena more attractive to private firms.

Yet this may not be enough to propel Africa to a clean, electrified future. Planners need more data to find the best sites for renewable energy projects. Developers are wary about pouring money into many countries, especially those with a history of corruption and governmental problems. And nations will need tens of billions of dollars to strengthen the energy infrastructure.

Still, green ambitions in Africa are higher now than ever before. Eddie O'Connor, chief executive of developer Mainstream Renewable Power in Dublin, sees great potential for renewable energy in Africa. His company is building solar and wind energy facilities there and he calls it "an unparalleled business opportunity for entrepreneurs."

Power Problems

Power outages are a common problem in many African nations, but Zambia has suffered more than most recently. It endured a string of frequent and long-lasting blackouts that crippled the economy. Pumps could not supply clean water to the capital, Lusaka, and industries had to slash production, leading to massive job layoffs.

The source of Zambia's energy woes was the worst drought in southern Africa in 35 years. The nation gets nearly 100% of its electricity from hydropower, mostly from three large dams, where water levels plummeted. Nearby Zimbabwe, South Africa and Botswana also had to curtail electricity production. And water shortages might get worse. Projections suggest that the warming climate could reduce rainfall in southern Africa even further in the second half of the twenty-first century.

Renewable energy could help to fill the gap, especially because wind and solar projects can be built much more quickly than hydropower, nuclear or fossil fuel plants. And green power installations can be expanded piecemeal as demand increases.

Egypt, Ethiopia, Kenya, Morocco and South Africa are leading the charge to build up renewable power, but one of the biggest barriers is insufficient data. Most existing maps of wind and solar resources in Africa do not contain enough detailed information to allow companies to select sites for projects, says Grace Wu, an energy researcher at the University of California, Berkeley. She co-authored a report[2] on planning renewable energy zones in 21 African countries, a joint project by the Lawrence Berkeley National Laboratory (LBNL) in California and the International Renewable Energy Agency (IRENA) in Abu Dhabi. The study is the most comprehensive mapping effort so far for most of those countries, says Wu. It weighs the amount of solar and wind energy in the nations, along with factors such as whether power projects would be close

"In the semiarid Karoo region of South Africa, a constellation of bright white wind turbines rises 150 meters above the rolling grassland."

to transmission infrastructure and customers, and whether they would cause social or environmental harm. "The IRENA–LBNL study is the only one that has applied a consistent methodology across a large region of Africa," says Wu.

High-resolution measurements of wind and solar resources have typically been done by government researchers or companies, which kept tight control of their data. The Berkeley team used a combination of satellite and ground measurements purchased from Vaisala, an environmental monitoring company based in Finland that has since made those data publicly available through IRENA's Global Atlas for Renewable Energy. The team also incorporated geospatial data—the locations of roads, towns, existing power lines and other factors—that could influence decisions about where to put energy projects.

"If there's a forest, you don't want to cut it down and put a solar plant there," says coauthor Ranjit Deshmukh, also an energy researcher at Berkeley.

The amount of green energy that could be harvested in Africa is absolutely massive, according to another IRENA report,[3] which synthesized 6 regional studies and found potential for 300 million megawatts of solar photovoltaic power and more than 250 million megawatts of wind. By contrast, the total installed generating capacity—the amount of electricity the entire continent could produce if all power plants were running at full tilt—was just 150,000 megawatts at the end of 2015. Solar and wind power accounted for only 3.6% of that.

The estimate of wind resources came as a surprise, says Oliver Knight, a senior energy specialist for the World Bank's Energy Sector Management Assistance Program in Washington DC. Although people have long been aware of Africa's solar potential, he says, as of about a decade ago, few local decision makers recognized the strength of the wind. "People would have told you there isn't any wind in regions such as East Africa."

The World Bank is doing its own studies, which assess wind speeds and solar radiation at least every 10 minutes at selected sites across target countries. It is asking governments to add their own geospatial data, and combining all the information into a user-friendly format that is freely available and doesn't require advanced technical knowledge, says Knight. "It should be possible for a midlevel civil servant in a developing country to get online and actually start playing with this."

South Africa Leads

In the semiarid Karoo region of South Africa, a constellation of bright white wind turbines rises 150 meters above the rolling grassland. Mainstream Renewable Power brought this project online in July 2016, 17 months after starting construction. The 35 turbines add 80 megawatts to South Africa's supply, enough to power about 70,000 homes there.

The Noupoort Wind Farm is just one of nearly 100 wind and solar projects that South Africa has developed since 2011, as the government launched a renewable energy accelerator program and prices fell below that of coal, causing construction to lag on two new massive coal plants. South Africa is primed to move quickly to expand renewable energy, in part thanks to its investment in data.

Environmental scientist Lydia Cape works for the Council for Scientific and Industrial Research, a national lab in Stellenbosch. She and her team have created planning maps for large-scale wind and solar development and grid expansion. Starting with data on the energy resources, they assessed possible development sites for many types of socioeconomic and environmental impacts, including proximity to electricity demand, economic benefits and effects on biodiversity.

The South African government accepted the team's recommendations and designated eight Renewable Energy Development Zones that are close to consumers and to transmission infrastructure—and where power projects would cause the least harm to people and ecosystems. They total "about 80,000 square kilometers, the size of Ireland or Scotland, roughly," says Cape. The areas have been given streamlined environmental authorization for renewable projects and transmission corridors, she says.

But for African nations to go green in a big way, they will need a huge influx of cash. Meeting sub-Saharan Africa's power needs will cost US$40.8 billion a year, equivalent to 6.35% of Africa's gross domestic product, according to the World Bank. Existing public funding falls far short, so attracting private investors is crucial. Yet many investors perceive African countries as risky, in part because agreements there require long and complex negotiations and capital costs are high. "It's a real challenge," says Daniel Kammen, who was a special envoy for energy for the US Department of State and is an energy researcher at the University of California, Berkeley. "Many of these countries have not had the best credit ratings."

Elham Ibrahim, the African Union's commissioner for infrastructure and energy, advises countries to take steps to reassure private investors. Clear legislation supporting renewable energy is key, she says, along with a track record of enforcing commercial laws.

South Africa is setting a good example. In 2011, it established a transparent process for project bidding called the Renewable Energy Independent Power Producer Procurement Programme (REIPPPP). The program has generated private investments of more than $14 billion to develop 6,327 megawatts of wind and solar.

Mainstream Renewable Power has won contracts for six wind farms and two solar photovoltaic plants through REIPPPP. "This program is purer than the driven snow," says O'Connor. "They publish their results. They give state guarantees. They don't delay you too much." Although the country's main electricity supplier has wavered in its support for renewables, the central government remains committed to the program, he says. "I would describe the risks in South Africa as far less than the risks in England in investing in renewables."

For countries less immediately attractive to investors, the World Bank Group launched the Scaling Solar project in January 2015. This reduces risk to investors with a suite of guarantees, says Yasser Charafi, principal investment officer for African infrastructure with the International Finance Corporation (IFC) in Dakar, which is part of the World Bank Group. Through the Scaling Solar program, the IFC offers low-priced loans; the World Bank guarantees that governments will buy the power generated by the projects; and the group's Multilateral Investment Guarantee Agency offers political insurance in case of a war or civil unrest.

Zambia, the first country to have access to Scaling Solar, won two solar projects that provided 73 megawatts and, in February 2017, signed a new agreement with IFC to develop up to an additional 500 megawatts. Madagascar, Senegal, and Ethiopia were next, with agreements to produce 30 to 40 megawatts, up to 200 megawatts, and up to 500 megawatts, respectively. IFC's target is to develop at least 1,000 megawatts in the first 5 years.

Making It Flow

Still, renewable power won't be useful if it can't get to users. One of the big barriers to a clean energy future in Africa is that the continent lacks robust electricity grids and transmission lines to move large amounts of power within countries and across regions.

But that gap also provides some opportunities. Without a lot of existing infrastructure and entrenched interests, countries there might be able to scale up renewable projects and manage electricity more nimbly than developed nations. That's what happened with the telephone industry: in the absence of much existing landline infrastructure, African nations rapidly embraced mobile phones.

The future could look very different from today's electricity industry. Experts say that Africa is likely to have a blend of power delivery options. Some consumers will get electricity from a grid, whereas people in rural areas and urban slums—where it is too remote or too expensive to connect to the grid—might end up with small-scale solar and wind installations and minigrids.

Still, grid-connected power is crucial for many city dwellers and for industrial development, says Ibrahim. And for renewables to become an important component of the energy landscape, the grid will need to be upgraded to handle fluctuations in solar and wind production. African nations can look to countries such as Germany and Denmark, which have pioneered ways to deal with the intermittent nature of renewable energy. One option is generating power with existing dams when solar and wind lag, and cutting hydropower when they are plentiful. Another technique shuttles electricity around the grid: for example, if solar drops off in one place, power generated by wind elsewhere can pick up the slack. A third strategy, called demand response, reduces electricity delivery to multiple customers by imperceptible amounts when demand is peaking.

These cutting-edge approaches require a smart grid and infrastructure that connects smaller grids in different regions so that they can share electricity. Africa has some of these "regional interconnections," but they are incomplete. Four planned major transmission corridors will need at least 16,500 kilometers of new transmission lines, costing more than $18 billion, says Ibrahim. Likewise, many countries' internal power grids are struggling to keep up.

That's part of what makes working in energy in Africa challenging. Prosper Amuquandoh is an inspector for the Ghana Energy Commission and the chief executive of Smart and Green Energy Group, an energy management firm in Accra. In Ghana, he says, "there's a lot of generation coming online."

The country plans to trade electricity with its neighbors in a West African Power Pool, Amuquandoh says, but the current grid cannot handle large amounts of intermittent power. Despite the challenges, he brims with enthusiasm when he talks about the future: "The prospects are huge!"

With prices of renewables falling, that kind of optimism is spreading across Africa. Electrifying the continent is a moral imperative for everyone, says Charafi. "We cannot just accept in the twenty-first century that hundreds of millions of people are left out."

Notes

[1] Africa Progress Panel. *Power, People, Planet: Seizing Africa's Energy and Climate Opportunities* (Africa Progress Panel, 2015).

[2] Wu, G. C., Deshmukh, R., Ndhlukula, K., Radojicic, T., and Reilly, J. *Renewable Energy Zones for the Africa Clean Energy Corridor* (IRENA/LBNL, 2015).

[3] Miketa, A., and Saadi, N. *Africa Power Sector: Planning and Prospects for Renewable Energy* (IRENA, 2015).

ANTHROPOCENE CITY: HOUSTON AS HYPEROBJECT (OR, WHEN THE NEXT HURRICANE HITS TEXAS)

Roy Scranton

(Portions of this essay first appeared in different form in The New York Times*, October 9, 2016, with the headline "Another Storm Is Coming." The complete essay appeared later in* Mustarinda*, cosponsored by the Finnish Cultural Institute in New York.)*

Imagine an oyster. Imagine waves of rain lashing concrete, a *crawdad boil, a fallen highway and a muddy bay. Imagine a complex system of gates and levees, the Johnson Space Center, a broken record on a broken player. Imagine the baroque intricacy of the Valero Houston oil refinery, the Petrobras Pasadena oil refinery, the LyondellBasell oil refinery, the Shell Deer Park oil refinery, the ExxonMobil Baytown oil refinery, a bottle of Ravishing Red nail polish, a glacier falling into the sea. Imagine gray-black clouds piling over the horizon, a chaos spiral hundreds of miles wide. Imagine a hurricane.*

Isaiah whirls through the sky, gathering strength from the Gulf of Mexico's warm waters. City, state, and federal officials do the sensible thing, evacuating beach towns and warning citizens and companies in Texas's petro-industrial enclaves from Bayou Vista to Morgan's Point to prepare for the worst.

The massive cyclone slows and intensifies as it nears the barrier islands off the coast, with wind speeds reaching over 150 mph. By sunset, several hours before landfall, the storm's counterclockwise arm is pushing water over the Galveston Seawall; by the time the eye finally crosses the beaches east of San Luis Pass, the historic city of Galveston has been flattened by a twenty-foot wave.

As Isaiah crosses into Galveston Bay, it only grows in strength, adding water to water, and when it hits the ExxonMobil Baytown refinery, some fifty miles inland, the storm surge is over twenty-five feet high. It crashes through refineries, chemical storage facilities, wharves and production plants all along the Houston Ship Channel, cleaving pipelines from their moorings, lifting and breaking storage tanks, and strewing toxic waste across east Houston.

The iridescent gray-brown flood rises, carrying jet fuel, sour crude, and natural gas liquids into strip malls, schools, and officces. By the time Isaiah passes inland, leaving the ruined coast behind, more than two hundred petrochemical storage tanks have been wrecked, more than a hundred million gallons of gas, oil, and other chemicals have been spilled, total economic damages for the region are estimated at over a hundred billion dollars, and three thousand six hundred eighty-two people have been killed. By most measures, it is one of the worst disasters in US history: worse than the 1906 San Francisco Earthquake, worse than Hurricane Katrina, worse than the terrorist attacks of September 11.

The effects ripple across the globe. The Gulf Coast is home to roughly 30 percent of the United States's proven oil reserves; The Gulf Coast and Texas hold 35 percent of its natural gas reserves. The refineries and plants circling Galveston Bay are responsible for roughly 25 percent of the United States's petroleum refining, more than 44 percent of its ethylene production, 40 percent of its specialty chemical feed stock and more than half of its jet fuel. Houston is the second busiest port in the United States in terms of pure tonnage and one of the most important storage and shipping points in the country for natural gas liquids. Isaiah shuts all that down. Within days of the hurricane's landfall, the NYSE and NASDAQ plummet as the price of oil skyrockets. Fuel shortages ground flights across the country, airline ticket prices soar, the prices of beef and pork shoot up, and gas prices at the pump leap to seven or eight dollars a gallon. The American economy slips into freefall.

Meanwhile, as the oil-poisoned water in east Houston flows back toward the sea, it leaves behind it the worst environmental catastrophe since the BP Deepwater Horizon spill. Rather than diffusing into open water, though, all the sludge is cradled within the protective arms of Galveston Bay.

The good news is that Isaiah hasn't happened. It's an imaginary calamity based on models and research. The bad news is that it's only a matter of time before it does. Any 50-mile stretch of the Texas coast can expect a hurricane once every six years on average, according to the National Weather Service. Only a few American cities are more vulnerable to hurricanes than Houston and Galveston, and not one of those is as crucial to the economy.

The worse news is that future hurricanes will actually be more severe than Isaiah. The models Isaiah is based on, developed by Rice University's Severe Storm Prediction, Education and Evacuation from Disaster (SSPEED) Center, don't account for climate change. According to Jim Blackburn, SSPEED's co-director, other models have shown much more alarming surges. "The City of Houston and FEMA did a climate change future," he told me, "and the surge in that scenario was 34 feet. Hurricanes are going to get bigger. No question. They are fueled by the heat of the ocean, and the ocean's warming. Our models are nowhere close."

Imagine Cobalt Yellow Lake. Imagine Cy Twombly's Say Goodbye, Catullus, to the Shores of Asia Minor. *Imagine colony collapse. Imagine refugees drowning off the shores of Asia Minor. Imagine causality, a bicycle tire, a million lost golf balls, a Styrofoam cooler, a bucket of crab claws, polyurethane, polypropylene, three copitas of mezcal, polyester, polyacrylic acid, polybutylene terephthalate, barbecue sauce, polycarbonate, polyether ether ketone, polyethylene, a Waffle House, polyoxymethylene, polyphenyl ether, polystyrene, the Wizard of Oz, polysulfone, polytetrafluoroethylene, polyvinyl chloride, a pair of pink crocs.*

I made a reservation aboard the M/V *Sam Houston* to take a boat tour of the Houston Ship Channel, the fifty-mile artery connecting Houston to the Gulf of Mexico, and the densest energy infrastructure nexus in North America. It seemed the perfect place to ask Timothy Morton about hyperobjects, dark ecology, and strange loops—some of the concepts he's been developing, as one of the leading thinkers of "speculative realism," in the effort to make philosophical sense of climate change.

The thinkers behind speculative realism, including Morton, Graham Harman, Quentin Meillassoux, and Jane Bennett, share a predilection for weird writers, woolly European metaphysics, and big ideas like the Anthropocene, but they'd likely resist being lumped all together. Graham Harman's "object-oriented ontology," for instance, argues that objects are autonomous in a way that keeps them from ever really connecting, perpetually withdrawing from each other in spite of apparent relations, while Jane Bennett's "vibrant matter" tells us that everything is equally alive and equally interwoven, humming together in a humongous, homogenous web in which a lost glove, an F-117 stealth bomber, and an Iraqi child are all basically the same kind of stuff. Morton, for his part, is more concerned with a critique of "Nature," arguing that we need to get past our cherished "culture/nature" divide in order to see ourselves as always already bound up in a dark mesh of ontological feedback.

As different as these thinkers are, though, they share a few key ideas. First, they all argue against what Meillassoux calls correlationism, the idea that human access to reality is limited to mere correlation between things-in-themselves and our thoughts about them. Our access to reality, they each insist in their own way, is more mysterious and complicated than just finding the circle-shaped thought for the circle-shaped thing. Second, for all these thinkers, things in the world have their own vitality independent of their relations to humans. A spoon has its own reality, as does an ocelot, a painting by Redon, or a Panamax container ship. Objects don't *need* human subjects to be meaningful, they argue, not even objects made by humans. Third, these thinkers all believe ontology trumps epistemology. Instead of asking how we can know things, that is, they insist we should be asking what it means for things to exist in the first place. The signature move that ties all this together is the willingness to indulge in speculative metaphysics—pondering what reality, deep down, really *is*. Spurning both mainstream analytic philosophy and the critical Marxist-Hegelian tradition, these thinkers have decided that what the world needs from philosophy isn't analysis, interpretation, or even transformation, but imagination.

Whether or not any of this makes any sense will depend on whom you ask. While speculative realism has generated a lot of buzz in literature departments and art magazines, its coherence and influence remain much debated. Some argue that object-oriented ontology is just a new way to fetishize commodities, especially the ones we call art. Others argue that the ideas behind speculative realism are specious and ignorant of the philosophical tradition. Climate scientists and academic philosophers, meanwhile, have hardly seemed to notice that speculative realism exists.

One of the reasons speculative realism exerts such a draw on artistic and literary types, I suspect, is because its thinkers make interesting aesthetic choices. This is especially true of Morton, who has a gift for the phrase. His book titles, capsule formulations of the ideas they elaborate, rumble with portent. Consider *Hyperobjects: Philosophy and Ecology after the End of the World*, *Ecology without Nature: Rethinking Environmental Aesthetics*, or *Dark Ecology: For a Logic of Future Coexistence*. Indeed, on the page Morton is a dizzying, acrobatic thinker; to read him is to take a wild ride through Romantic poetry, Western philosophy, literary theory, and climate change—imagine Slavoj Žižek on psilocybin.

In person, Morton is gentle, funny, and self-effacing, equal parts Oxbridge and cybergoth. We drove out to the ship channel in his white Mazda. As we rose and fell through the soaring grandeur of Houston's swooping highway exchanges, we talked about writing practice and work-life balance: Morton had two books coming out in 2016 and was writing two more, and when he's not busy writing, spending time with his kids, giving lectures, blogging, or collaborating with Björk, he teaches courses on literary theory and "Arts in the Anthropocene" at Rice University, where he holds the Rita Shea Guffey Chair in English.

Turning off the highway, we descended into the petro-industrial gray zone that sprawls from Houston to the sea. A Port of Houston security guard checked our IDs and we drove past hundred-foot-long turbine blades, massive shafts, and what looked like pieces of giant disassembled robots. I asked Tim how he liked living in Houston.

"This is the dirty coast," he said. "Dirty in the sense that something's wrong. We're holding this horrible, necessary energy substance, and it's like working in an emergency room or a graveyard or a charnel ground. You're basically working with corpses, with fossils from millions of years ago, you're working with deadly toxic stuff all the time, stuff that has very intense emotion connected to it. If I was going to find a word that described Texan-ness, I'd use the word 'wild'—phenomenologically, emotionally, experientially wild."

We parked and boarded the M/V *Sam Houston*. As the boat spun away from the pier and headed east, Tim and I went out on deck. Across the brown-black water enormous claws and magnets shifted scrap metal from one heap to another, throwing up clouds of metal dust, while the engine thrummed through my feet and the wind whipped across the mike of my voice recorder.

"The thing is," Tim said, "being aware of ecological facts is the very opposite of thinking about or looking at or talking about nature. *Nature* is always conceptualized as an entity that's different or distinct from me somehow. It's in my DNA, it's under my clothes, it's under the floorboards, it's in the wilderness. It's everywhere except for right here. But ecology means it's in your face. It *is* your face. It's part of you and you're part of it."

Several industrial recycling companies line the upper reaches of the Houston Ship Channel, including Derichebourg Recycling USA, Texas Port Recycling, and Cronimet USA, all recognized emitters of one of the most potent carcinogens known to science, hexavalent chromium. Behind the giant cranes and heaps of scrap lies the predominantly Hispanic neighborhood of Magnolia Park, whose residents have long complained of unexplained smoke and gas emissions, persistent pollution, and strange, multi-colored explosions.

"The simplest way of describing that is *ecology without nature*," Tim continued. "That doesn't mean I don't believe in things like coral. I believe in coral much more than someone who thinks that coral is this 'natural' thing. Coral is a life form that's connected to other forms. Everything's connected. And how we think about stuff is connected to the stuff. How you think about stuff, how you perceive stuff, is entangled with what you're perceiving."

In among the recycling yards sat Brady's Landing, a steak-and-shrimp restaurant. Through its plate-glass windows dozens of empty white tables shone like pearls in black velvet. I imagined diners eating crab-stuffed trout, watching the water rise up over the Ceres wharfs across the channel, rise up over the pilings at the edge of Brady's Island, rise up over the restaurant's foundations and up the windows, one foot, two feet, six feet, and the glass would crack, creak, and burst open, and the tide would rush in over fine leather shoes and French cuffs and napkin-covered laps and lift them, the diners, their tables, plates, pinot noir, and crab-stuffed trout, lift them and spin them in a rich and strange ballet.

"It's like when you realize you're actually a life form," Tim said. "I'm Tim but I'm also a human. That sounds obvious but it isn't. I'm Tim but I've also got these bits of fish and viral material inside me, that *are* me. That's not a nice, cozy experience; it's an uncanny, weird experience. But there's a kind of smile from that experience,

"At some point, instead of trying
to delete the twisty darkness, you
have to make friends with it. And
when you make friends with it,
it becomes strangely sweet."

because ecological reality is like that. Ecological phenomena are all about loops, feedback loops, and this very tragic loop we're on where we're destroying Earth as we know it."

Interstate Highway 610 loomed above, eighteen-wheelers and SUVs rolling through the sky. In the distance, gas flares flashed against the cloud cover. Pipes fed into pipes that wrapped back into pipes circling pipes, Escher machines in aluminum and steel.

"Ecological thinking is about never being able to be completely in the center of your world. It's about everything seeming out of place and unreal. That's the feel of dark ecology. But it isn't just about human awareness: it's about how *everything* has this uncanny, looped quality to it. It's actually part of how things are. So it's about being horrified and upset and traumatized and shocked by what we've been up to as human beings, and it's about realizing that this basic feeling of twistedness isn't going away."

A voice boomed out from the bowels of the boat as we broke from the highway's shadow: "First refinery to the right is Valero. This refinery began operations in 1942. It will handle 145,000 barrels of oil per day." Directly behind Valero lay Hartman Park, with its green lawns and baseball diamonds—the jewel of Manchester, one of the most polluted neighborhoods in the United States. Manchester is blocked in on the north by Valero, and on the east, south, and west by a chemical plant, a car-crushing yard, a water treatment plant, a train yard, Interstate 610, and a Goodyear synthetic rubber plant. In 2010, the EPA found toxic levels of seven different carcinogens in the neighborhood. The area is 88% Hispanic.

"At some point," Tim said, "instead of trying to delete the twisty darkness, you have to make friends with it. And when you make friends with it, it becomes strangely sweet."

Imagine Greenland. Imagine Kellogg, Brown, & Root. Imagine Uber, the Svalbard Seed Vault, a roadkill raccoon, six months in juvie, Green Revolution, amnesia. Imagine ZZ Top. Imagine White Oak Bayou flooding its banks. Imagine Mexican gardeners swinging weed-eaters. Imagine boom and bust, the murmur of Diane Rehm, sizzurp, a sick coot, Juneteenth, coral bleaching, amnesia. Imagine losing Shanghai, New York, and Mumbai. Imagine "In the Mood." Imagine amnesia.

From Houston, the ship channel goes south through Galveston Bay, cutting a trench approximately 530 feet wide and 45 feet deep through the estuary bottom to where it passes into the Gulf of Mexico. As you follow the channel south along I-45, strip clubs and fast food franchises give way to bayou resorts and refineries, until the highway finally leaps into the air, soaring over the water with the pelicans. It comes down again in downtown Galveston, once known as "Wall Street of the South": a mix of historic homes, dry-docked oil rigs, beach bars, and the University of Texas Medical Branch. The gulf spreads sullen and muddy to the south, its placid skin broken by distant blisters of flaming steel.

Galveston Bay is a Texas paradox. One of the most productive estuaries in the United States, it offers up huge catches of shrimp, blue crab, oysters, croaker, flounder and catfish, and supports dozens of other kinds of fish, turtles, dolphins, salamanders, sharks and snakes, as well as hundreds of species of birds. Yet the bay is heavily polluted, so full of P.C.B.s, pesticides, dioxin and petrochemicals that fishing is widely restricted. The bay is Houston's shield, protecting it from the worst of the Gulf Coast's weather by absorbing storm surges and soaking up rainfall, but hydrologists at Rice University are worried it might also be Houston's doom: The wide, shallow basin could, under the right conditions, supercharge a storm surge right up the ship channel.

The fight to protect Houston and Galveston from storms has been going on for more than a century, ever since Galveston built a 17-foot sea wall after the Great Storm of 1900, a Category 4 hurricane that killed an estimated 10,000 to 12,000 people. The fight has been mainly reactive, always planning for the last big storm, rarely for the next. The levees around Texas City, for instance, were built after Hurricane Carla submerged the chemical plants there in 10 feet of water in 1961. Today, Hurricane Ike, which hit Texas in 2008, offers the object lesson.

Hurricane Ike was a lucky hit with unlucky timing. Forecasts had the hurricane landing at the southern end of Galveston Island, and if they'd been right, Ike would have looked a lot like Isaiah. Instead, in the early morning hours of Sept. 13, 2008, Ike bent north and hit Galveston dead on, which shifted the most damaging winds east. The sparsely populated Bolivar Peninsula was flattened, but Houston came out okay.

Still, Ike killed nearly 50 people in Texas alone, left thousands homeless, and was the third costliest hurricane in American history. It would have been the ideal moment for Texas to ask Congress to fund a comprehensive coastal protection system. But on that Monday, Sept. 15, Lehman Brothers filed the largest bankruptcy in United States history, and the next day the Federal Reserve stepped in to save the failing insurance behemoth A.I.G. with an $85 billion bailout. Nature's fury took a back seat to the crisis of capital.

Since then, two main research teams have led the way in preparing for the next big storm: Bill Merrell's "Ike Dike" team at Texas A&M Galveston (TAMUG), and the SSPEED Center at Rice University, led by Phil Bedient and Jim Blackburn. Despite shared goals, though, the relationship between the two teams hasn't always been easy. Bill Merrell's cantankerous personality and obsessive drive to protect Galveston have clashed with SSPEED's complex, interdisciplinary, Houston-centric approach.

Dr. Merrell's Ike Dike has the blessing of simplicity, which softens the sticker shock: It is estimated to cost between $6 billion and $13 billion. The plan is to build a 55-mile-long "coastal spine" along the gulf. The plan's main disadvantage is that a strong enough hurricane could still flood the Houston Ship Channel, because of what Dr. Bedient calls the "Lake Okeechobee effect."

"The Okeechobee hurricane came into Florida in 1928 and sloshed water to a 20-foot surge," Dr. Bedient explained. "Killed 2,000 people. But Lake Okeechobee is unconnected to the coast. It was just wind. Galveston Bay has the same dimensions and depth as Lake Okeechobee in Florida. So imagine we block off Galveston Bay with a coastal spine, and we have a Lake Okeechobee."

Dr. Bedient worked on the Murphy's Oil spill in St. Bernard Parish, La., where flooding from Hurricane Katrina ruptured a storage tank, releasing more than a million gallons of oil and ruining approximately 1,800 homes. One of Dr. Bedient's biggest worries is what a storm might do to the estimated 4,500 similar tanks surrounding Houston, many of them along the ship channel. If even 2 percent of those tanks were to fail because of storm surge, the results would be catastrophic.

The SSPEED Center advocates a layered defense, including a mid-bay gate that could be closed during a storm to protect the channel. On its face, the plan seems unwieldy, but SSPEED's models show it could stop most of the surge from going up the ship channel, with or without the Ike Dike, at an estimated cost of only a few billion dollars.

On the government side, various entities are at work in the ponderous and opaque way of American bureaucracy. The US Army Corps of Engineers has its own research and development process, and is working on a study of the Galveston-Houston area as part of its more comprehensive Gulf Coast research agenda, which could, eventually, lead to a recommendation for further studies, feasibility and cost-benefit analyses, environmental impact reports, and perhaps someday a project, which, were it funded by Congress, might even get built. One must be patient. It took the Army Corps of Engineers twenty-six years to build the Texas City Levee. When Katrina hit New Orleans and breached the levee system there, the Corps had been working on it since 1965, and it was still under construction.

Meanwhile, the Gulf Coast Community Protection and Recovery District is working to synthesize SSPEED and TAMUG's work into its own proposal. The GCCPRD was established by Texas governor Rick Perry in 2009, in the wake of Ike, but wasn't funded until 2013, when the Texas General Land Office stepped in with a federal grant from the Department of Housing and Urban Development. The GCCPRD board comprises county judges from Brazoria, Chambers, Galveston, Harris, Jefferson, and Orange counties, three additional members, and a President, currently former Harris County judge Robert Eckels, and has hired Dannenbaum Engineering, a local company with a strong track record in public infrastructure, to put the report together. The GCCPRD takes its lead from the GLO, headed today by Commissioner George P. Bush, and the specific language of the HUD grant restricts their work to analysis and general-level planning. Any more specific plans will have to come later, pending additional funding.

If there's one thing Houston can teach us about the Anthropocene, it's that all global warming is local. I went down myself to see representatives from all of these organizations—the USACE, SSPEED, TAMUG, the GLO, and the GCCPRD, plus the Texas Chemical Council and the Bay Area Houston Economic Partnership—testify before the State of Texas Joint Interim Committee on Coastal Barrier Systems (JICCBS), a special committee of the Texas state legislature, held at the TAMUG campus in Galveston.

Over five hours of presentations, talking points, and questions, a rough sense of the future began to take shape. As I sat in the back row listening to politicians ask about how various projects might affect insurance rates, how long different projects might take to build, and how the pitch could be put to the US Congress asking for the billions of dollars needed, I imagined a single white feather, numinous in the golden light of the PowerPoint, drifting across the conference room, floating over the heads of the senators, administrators, and scientists, and rising, rising, rising on an ever-expanding wave of confidence.

What obstacles might have remained between this roomful of committed public servants and the building of one of the largest coastal infrastructure projects in the world seemed for a moment insubstantial. The fact that environmental impact studies taking years to complete had yet to be started, that any of the land in question would have to be bought or seized under eminent domain, that all the planning at this stage was merely notional and actual designs would have to be bid on, contracted out, and approved, that there was no governmental agency in place to take responsibility for a coastal barrier system and maintain it, much less build it, and that somebody still had to come up with the money, somewhere, perhaps somehow convincing divided Republicans and embattled Democrats in the US Congress to send a bunch of Texas pols and their cronies a check for $13 billion—these were all mere details, nothing to worry about. I felt sure the political will manifest in that conference room would find a way.

And I had total confidence that those same feelings of goodwill, pragmatism, and accomplishment would be found, more or less, at the next Joint Interim Committee on Coastal Barrier Systems meeting, and the next academic conference on "Avoiding Disaster," and the next policy symposium on energy transition, and the next global conference on sea level rise, and the next plenary on carbon trading, and the next colloquium on the Anthropocene, and the next Conference of Parties to the United Nations Framework Convention on Climate Change, and the next, and the next, and the next, and the next, and journalists would report on it, and philosophers would ponder it, and activists would tweet about it, and concerned people like you would read about it. The problem is, it's not enough.

According to Jim Blackburn, "Even a locally funded project would probably be three years in the permitting and another six to eight years in construction." Most local politicians, however, seemed to prefer the Ike Dike, necessarily a Federal project. "I have heard more than one person say our plan is to wait until the next hurricane comes," Blackburn said, "and then depend on guilt money from Washington to fix the problem."

Bill Merrell told me much the same thing: "We see local politicians in general content with doing nothing. The do-nothing option is pretty gruesome. It gets you a storm, sooner or later, that's going to kill thousands of people and cause at least $100 billion in damage. The cost of doing nothing is horrendous. But trying to get politicians from doing nothing to doing something is really hard. I think I've started to appreciate that more. I didn't realize it would be as hard as it was."

Two weeks after the JICCBS meeting, Houston was inundated with more than a foot of rain in less than twenty-four hours, almost two feet in some neighborhoods. Flooding damaged more than 200 homes and killed eight people. By the end of the month, it was the wettest April the city had ever recorded. More rain and more floods hit Texas in May and June, then a week of precipitation in August dumped around seven trillion gallons of water on Louisiana, with some areas accumulating more than twenty inches of rain. Flooding killed 13 people and damaged 146,000 homes.

Imagine Earth. Imagine "Pretty Hurts." Imagine Lakewood Church, wind-lashed magnolias, a bottle of Topo Chico, the Astrodome. Imagine surface and depth, weather drones, the Geto Boys, thermodynamic disequilibrium, a body in a hole. Imagine the economy slowing, snowy egrets nesting in a live oak, becoming one with the Ocean of Soul, a Colt Expanse carbine. Imagine Purple Drank and a bowl of queso. Imagine Terms of Endearment. *Imagine stocks and flows, a pearl, a rhizome. Imagine the end of the world as we know it.*

The *Sam Houston*'s ninety-minute tour of the Houston Ship Channel only goes a few miles out before turning around at the LyondellBasell refinery, one of the largest heavy-sulfur-crude refineries in the US, processing around 268,000 barrels a day. The loudspeaker voice offered us complimentary soft drinks. I asked Tim Morton whether dark ecology had a politics.

"Obviously," he said, "it's not just that unequal distribution is connected to ecological stuff. It *is* ecological. It's not like we need to condescend to include fighting racism and these other issues under the banner of ecological thinking. It's the other way around. These problems were *already* ecological because the class system is a Mesopotamian construct and we're basically living in Mesopotamia 9.0. We're looking at these oil refineries and it's basically an upgrade of an upgrade of an upgrade of an agricultural logistics that began around 10,000 BC and is directly responsible, right now, for a huge amount of carbon emissions but also absolutely necessitated industry and therefore global warming and mass extinction."

We passed the CEMEX Houston Cement Company East Plant, the Gulf Coast Waste Disposal Authority's Washburn Tunnel Wastewater Treatment Facility, the Kinder Morgan Terminal, and Calpine's Channel Energy Center, a natural gas steam plant.

"This is where I have to say something English, which is 'Give us a chance, mate.' Because we can't do everything all at once, and we come to the conversation with the limitations and the skill sets that we have, and we're getting round to stuff. But maybe the first thing to do is to notice: We. Are. In. A. Shit. Situation. Maybe the first thing to do is go, okay, we're causing a mass extinction the likes of which hasn't been seen since the end-Permian extinction that wiped out 95% of life on earth. Dark ecology has a politics but it's a very different kind of politics because it means that the idea that humans get to decide what reality is needs to be dismantled. It's an ontological war."

Off our starboard, Public Grain Elevator #2 poured wheat into the hold of a Chinese freighter, a hundred yards from a giant mound of yellow Mexican gypsum. The Valero refinery rose again to port, flare stacks burning against the sky, just beyond where Sims Bayou broke off from the channel and meandered in toward South Park and Sunnyside, poverty-stricken African-American neighborhoods largely abandoned by Houston's government. One area of Sunnyside was recently rated the second most dangerous neighborhood in America. 76% of the children there live in poverty. Residents have a 1 in 11 chance of becoming a victim of violent crime.

"Take hyperobjects," Tim said, staring fixedly at the Valero refinery. "Hyperobjects are things that are so huge and so long-lasting that you can't point to them directly, you can only point to symptoms or parts of them. You can only point to little slivers of how they appear in your world. Imagine all the oil on earth, forever, and the consequences of extracting and burning it for the next

100,000 years. That would be a hyperobject. We're going through this ship channel and these huge gigantic entities are all symptoms of this even larger, much more disturbing thing that we can't point to directly. You're in it and you *are* it, and you can't say where it starts and where it stops. Nevertheless, it's this thing here, it's on Earth, we know where it is."

We passed Brady's Landing and Derichebourg Recycling and Bray's Bayou. The boat motored back much faster than it had gone out, and I had to strain to catch Tim's voice against the noise of the wind and water.

"My whole body's full of oil products," he said. "I'm wearing them and I'm driving them and I'm talking about them and I'm ignoring them and I'm pouring them into my gas tank, all these things I'm doing with them, precisely *that* is why I can't grasp them. It's not an abstraction. It's actually so real that I can't point to it. The human species is like that: instead of being this thing underneath appearance that you can point to, it's this incredibly distributed thing that you *can't* point to. The one thing that we need to be thinking right now, which is that *as a human being* I'm responsible for global warming, is actually quite tricky to fully conceptualize."

The *Sam Houston* throttled down and bumped against the wharf, the crew laid out the gangplank, and we disembarked. Tim and I got back in his white Mazda and he punched directions into his phone.

"Make a U-turn," the fembot voice commanded. We drove out past the guard shack and over some railroad tracks, then out onto the highway.

"Anybody who's got any intelligence or sensitivity working with this stuff very quickly gets into dilemma space," Tim said, changing lanes. "I think it's a matter of nuance, how you work with that. I admire any mode of thought that goes as quickly as possible to this dilemma space, but we've only just begun to notice the 'we' doing these horrible things, and it's okay to be completely confused and upset. We're in shock, and that's on a good day. Most days it's just grief work because we're in a state of total denial. I am too. I can only allow myself to feel really upset about what's going on for maybe one second a day, otherwise I'd be in a heap on the floor all the time crying."

We took Alt-90 to I-10, passing a Chevron and a Shell and a Subway and Tires R Us and Mucho Mexico, then rose into the flow of traffic cruising the interstate west.

"We're constantly trying to get on top of whatever we're worrying about, but if you look at it from an Earth magnitude, that's magical thinking. We've given ourselves an impossible-to-solve problem. The way in which we think about the problem, the way in which we give it to ourselves, is part of the problem. How do you talk to people in a deep state of grief when you're also in that deep state of grief?"

The lanes split and we wove from I-10 to 59 and then, just past Fiesta Mart's enormous neon parrot, slid down the ramp to Fannin Street.

"I think there's an exit route, actually," Tim said, "but it's paradoxical. It involves going down underneath: it's not about transcending in any sense, it's about what I call subscending. There's always so much more about weather than just being a symptom of global warming. It *is* a symptom of global warming, but it's also a bath, these little birds over here, it's this wonderful wetness on the back of my neck, it's this irritating thing that's clogging up my drain."

We passed Fannin Flowers and then turned onto Bissonnet Street, rolling by the Museum of Fine Arts, with its special exhibit on Art Deco cars, and the Contemporary Art Museum, which was featuring an exhibition about the colonization of Mars. We turned in past Mel Chin's *Manilla Palm*, a giant fiberglass and burlap tree erupting out of a broken steel pyramid, then turned again, tracking back toward my apartment, past expensive new condos and down the dead end where I lived. Bamboo rose against the fence at the end of the road. Tim parked by the curb and shut off the Mazda.

"It boils down to knowing that global warming is a catastrophe rather than a disaster. Disasters are things that you rubberneck as they're happening to other people because you're reading about it in the Book of Revelation. It isn't now. A catastrophe, on the other hand, is something that you're inside of and it's got this weird, loopy, twisty structure to it. Disaster's like: everything's being destroyed and I can see perfectly how everything's being destroyed. Catastrophe's more like: OMG, I am the destruction. I'm part of it and I'm in it and I'm on it. It's an aesthetic experience, I'm inside it, I'm involved, I'm implicated."

A cardinal flew across the street, a streak of red against the green.

"I think that's how we get to smile, eventually, by fully inhabiting catastrophe space, in the same way that eventually a nightmare can become so horrible that you start laughing. That's how you find the exit route. I feel like maybe part of my job is giving people that."

Imagine black. Imagine black, black, black, blue-black, red-black, purple-black, gray-black, black on black. Imagine methane. Imagine education. Imagine wetlands. Imagine a brown-skinned woman in white circling the Rothko Chapel chanting "Zong. Zong. Zong." Imagine a regional, comprehensive approach to storm surge risk management, lemonade, the Slab Parade, increased capacity, complexity, attribution studies, progress, a wine and cheese reception, TACC's Stampede supercomputer, an integrative, place-based research program, Venice's Piazza San Marco, sea level rise, Destiny's Child. Imagine a red line. Imagine two degrees. Imagine flare stacks. Imagine death.

Maybe it was the eleventh straight month of record-breaking warming. Maybe it was when the Earth's temperature hit 1.5 degrees Celsius over pre-industrial levels. Maybe it was new reports that Antarctica and the Arctic were melting faster than anyone expected. Maybe it was when Greenland started melting two months early, and then so quickly that scientists didn't believe their data.

Maybe it was watching our world start to come apart, and knowing that nothing would be done until it was too late.

We've known that climate change was a threat since at least 1988, and the United States has done almost nothing to stop it. Today it might be too late. The feedback mechanisms that scientists have warned us about are happening. Our world is changing.

Imagine we've got twenty or thirty years before things really get bad. Imagine how that happens. Imagine soldiers putting you on a bus, imagine nine months in a FEMA trailer, imagine nine years in a temporary camp. Imagine watching the rich on the other side of the fence, the ones who can afford beef and gasoline, the ones who can afford clean water. Imagine your child growing up never knowing satiety, never knowing comfort, never knowing snow. Imagine politics in a world on fire.

Climate change is hard to think about not only because it's complex and politically contentious, not only because it's cognitively almost impossible to keep in mind the intricate relationships that tie together an oil well in Venezuela, Siberian permafrost, Saudi F-15s bombing a Yemeni wedding, subsidence along the Jersey Shore, albedo effect near Kangerlussuaq, the Pacific Decadal Oscillation, the polar vortex, shampoo, California cattle, the Great Pacific Garbage Patch, leukemia, plastic, paper, the Sixth Extinction, Zika, and the basic decisions we make every day, are forced to make every day, in a world we didn't choose but were thrown into. No, it's not just because it's mind-bendingly difficult to connect the dots. Climate change is hard to think about because it's depressing and scary.

Thinking seriously about climate change forces us to face the fact that nobody's driving the car, nobody's in charge, nobody knows how to "fix it." And even if we had a driver, there's a bigger problem: no car. There's no mechanism for uniting the entire human species to move together in one direction. There are more than seven billion of us, and we divide into almost two hundred nations, thousands of smaller sub-national states, territories, counties, and municipalities, and an unimaginable multitude of corporations, community organizations, neighborhoods, religious sects, ethnic identities, clans, tribes, gangs, clubs, and families, each of which faces its own internal conflicts, disunion, and strife, all the way down to the individual human soul in conflict with itself, torn between fear and desire, hard sacrifice and easy cruelty, all of us improvising day by day, moment by moment, making decisions based on best guesses, gut hunches, comforting illusions, and too little data.

But that's the human way: reactive, ad hoc, improvised. Our ability to reconfigure our collective existence in response to changing environmental conditions has been our greatest adaptive trait. Unfortunately for us, we're still not very good at controlling the future. What we're good at is telling ourselves the stories we want to hear, the stories that help us cope with existence in an wild, unpredictable world.

Imagine life. Imagine a hurricane. Imagine a brown-skinned woman in white circling the Rothko Chapel chanting "Zong." Imagine grief. Imagine the Greenland ice sheet collapsing and black-crowned night herons nesting in the Live Oaks. Imagine Cy Twombly's Say Goodbye, Catullus, to the Shores of Asia Minor, amnesia, a broken record on a broken player, a tar-stained bird, the baroque complexity of a flooded oil refinery, glaciers sliding into the sea. Imagine an oyster. Imagine gray-black clouds piling over the horizon, a sublime spiral hundreds of miles wide. Imagine climate change. Imagine a happy ending.

HAUNTINGS IN THE ANTHROPOCENE: AN INITIAL EXPLORATION

Jeff VanderMeer

(First appeared in Environmental Critique, *July 7, 2016)*

1.

Timothy Morton's Hyperobjects, which sets out a series of thoughts about "dark ecologies," has become central to thinking about storytelling in the modern era, in my opinion. Morton's central idea of a hyperobject is in a sense a way of using a word as an anchor for something that would be otherwise hard to picture in its entirety–it is an all-encompassing metaphor that also has its own reality, both literal and figurative, here and there. The word therefore is a very important signifier for any fiction writer wishing to engage with the fragmented and diffuse issues related to the Anthropocene.

What is a hyperobject? Something viscous (they stick—to your mind, to the environment) and nonlocal (local versions are manifestations from afar). Their unique temporality renders them invisible to human beings for stretches of time and they exhibit effects in the interrelationship of objects. In the instance of most interest to both Morton and to me, global warming can be considered as hyperobject. And even with just this bare context given, it should

"The uncanny has infiltrated the real,
and in some sense that boundary
is forever compromised."

be clear why the term is of use. Because a hyperobject is everywhere and nowhere, cannot really be held in one place by the human brain, reaction to it by the human world is often irrational or inefficient or wrong.

If global warming in the Anthropocene can be identified in general as a hyperobject, there is perhaps further value in describing it specifically as a kind of haunting. At least, this is of extreme interest to me as a fiction writer and someone who wants to find new ways of telling stories that better fit the extremes of our era.

As Maria de Pilar Blanco and Esther Peeren write in "Possessions: Spectral Places" in *The Spectralities Reader* (Bloomsbury), "Repetition of events, images, and localities is one of the recurrent motifs of the uncanny." What is global warming but repetitions bound by laws of cause and effect that come to feel uncanny because no one can see the entire outline of a hyperobject (i.e., elements of both cause and effect seem invisible)? Global warming is an inherently destabilizing force for this reason, whereas the uncanny is neutral because it can be used to either destabilize the reader's perception of the world, or, by story's end, to reinforce the status quo.

Rather than creating escapism, mapping elements of the Anthropocene via weird fiction may create a greater and more visceral understanding (render *more* visible)—precisely because so many of the effects of this era are felt in and under the skin, as well as in the subconscious (whether manifesting as a denial of civilization's death or in a more personal manner).

In the introduction to the anthology *The Weird*, coedited with my wife Ann VanderMeer, I wrote that the weird tale "represents the pursuit of some indefinable and perhaps maddeningly unreachable understanding of the world beyond the mundane—a 'certain atmosphere of breathless and unexplainable dread' or 'particular suspension or defeat of…fixed laws of Nature.'"*

In the modern era, the hyperobject of global warming makes such a mockery of what our five senses can perceive that the "fixed laws of Nature" seem more and more, through, for example, extreme weather events, to have become un-fixed, the compass spinning wildly. The laws of science, which often seem resolute, begin to seem less so, even if this is just our faulty perception.

The uncanny has infiltrated the real, and in some sense that boundary is forever compromised. What, for example, is William T. Vollmann's *Imperial*, a thousand-page rumination on the psycho-ecological cost of the ruination of the Salton Sea, other than a vast and apocalyptic haunting?

It is a haunting first of the author, who confesses at one point he could not remove his personal perspective because he was too affected by what he was observing. It is a haunting second in the reader through the repetition of description of ecological impact and loss, which begins to manifest as physical stress or nightmares. The book is an especially potent example of purgatory in the uncanny, because it provides a glimpse of mid-Collapse in the Anthropocene—a transitional state in which those affected may not even realize the progression of decay *in the moment*. This kind of haunting in progress dislocates and relocates both the mind and the body. The value of such a book lies not in its facts or its adherence to ideas about science, but in conveying the totality of this mid-Collapse condition.

Extrapolating outward from the epicenter of the Salton Sea, we find the visible in the invisible and the same repetition across other bodies of water and water in general—which has become the ultimate uncanny element. The consequences of our actions in even the deepest parts of the ocean lie hidden from us, "out of sight, out of mind." Plastic bottles of water are also part of the visible invisible, the repetition of idea and ideology we don't often acknowledge or don't know what to do with—a bottle of water is at this point life and death packaged in the same object. We, in our millions, like a cute octopus hiding in an open soda can on the sea bed, surrounded by a desolate landscape denuded of life without acknowledging we're looking at trash. Only when the evidence is too visible or extraordinary to be overlooked or subsumed by the landscape does this kind of haunting become noticeable—as when toxic algae bloom on the coast of Florida in part because of the untenable practices of large-scale agriculture. Otherwise, it is an unpleasant hum in the background that continues to colonize our bodies and our lives despite our inattention to it. Even as the pumps being built on Miami Beach to keep the water out make sounds as they work like a dark absurdist laughter from an unseen specter.

2.

Reading about hauntings and thinking about them in terms of Vollmann's book helped me understand where the idea for Area X in my Southern Reach trilogy came from…something that was not immediately apparent. Yes, I had a dream of walking down into a tunnel-tower and seeing living words on the walls. Yes, I woke up and the biologist character was in my head and that next morning I wrote the first ten pages of *Annihilation*, which would remain more or less unchanged in the final draft.

But for a long time I didn't realize what irritant or issue or problem had lodged in my subconscious to force Area X out. Finally, though, I realized that the Gulf Oil Spill had created Area X. It wasn't something I said out loud at first because it sounds vaguely or not so vaguely pretentious—overly earnest. But it's true. By the time of the Gulf Oil Spill I had lived in North Florida for over 20 years and through my hiking and experiences in the wilderness along that coast I felt for the first time in a wandering life like I belonged in a place, in a landscape.

Then suddenly the oil was gushing out in the Gulf, and it couldn't be contained, and for many of us in the area it was gushing in our minds, and we could not get away from it. It was haunting us day and night, always there—a phantom sound, a phantom thought. For a time, for more than a month it wasn't clear the well would ever be capped. For more than a month, many of us thought this catastrophe would last for years, and the Gulf and Gulf Coast would be, in essence, gone.

But even after they capped the well, it was still somewhere in the back of my mind, and eventually that dark swirl coalesced into a dark tunnel with words on the wall, and an invisible border and Area X: a strange place in which nature was always becoming more what it had always been without human interference: less contaminated, less compromised. Safe. Where the oil was being taken out.

Oil, like water, creates a particular kind of haunting. The best recent example might be the airport scene in Tom McCarthy's *Satin Island*. The narrator fixates on the scenes of oil that haunt the backdrop of news reports as he walks through the transitional space of the terminal and the details about oil not only place it in the foreground, leaking out over fellow travelers but, in the descriptions, oil attains a kind of agency or power—and an intentional fetishizing—that is, indeed, almost uncanny. McCarthy's hierarchy in the scene inhabits both the surface and subtext, providing a view of oil as a more ubiquitous and over-arching definition of planet Earth than humankind. After reading this scene, it would be impossible to view oil as "inert" or unalive ever again, or as non-political. Supposedly we already know these things, but sometimes fiction can make us feel them in our bones.

3.

The uncanny has infiltrated the real, and in some sense that boundary is forever compromised. There may be only one way into Area X, but there are a thousand ways out. As Michel de Certeau explains in *The Practice of Everyday Life*, "Every place has its own...proliferation of stories and every spatial practice constitutes a form of re-narrating or re-writing a place...Walking [into a place] affirms, suspects, tries out, transgresses, respects...haunted places are the only ones people can live in."

In the second Southern Reach novel, *Authority*, the hauntings are smaller and more intimate. The size of rooms inside the Southern Reach building often doesn't match their insides and lines of dialogue from *Annihilation* spirit across the hallways the main character

Control walks down and a member of the twelfth expedition comes back as someone called Ghost Bird and the psychological overlays of some of the characters, across both the landscape and fellow employees, form a haunting with multiple sedimentary levels. The obsolete tech of past decades haunts the present.

With that sanctuary left behind in *Acceptance*, manifestations pour out across the sky, birds "trailing blurs of color that resemble other versions of themselves and the air seemed malleable, or like it could be convinced or coerced...In the lengthening silence and solitude, Area X sometimes would reveal itself in unexpected ways. Standing in a clearing one evening, I felt a kind of breath or thickness of molecules behind that I could not identify, and I willed my heartbeat to slow, hoping to be so quiet that without turning I might hear or in some other way glimpse what regarded me. But to my relief it fled or withdrew into the ground a moment later."

"The correlation between movement and progress is broken [in a haunting] and progress is broken and the subject succumbs to a feeling of ungroundedness and spatio-temporal disjointedness," Blanco and Peeren write.

"Time runs on time," writes the great dark ecologies poet Aase Berg, "Time runs on time and starvation and the weakness carries me in across the gray regions. And the soul's dark night will slowly be lowered through me. That is why I now slowly fold myself like a muscle against the wet clay...I will sleep now in my bird's body in the down, and a bitter star will radiate eternally above the glowing face's watercourse."

Thomas Hardy writes of a fallen soldier, "Yet portion of that unknown plain / Will Hodge for ever be; / His homely Northern breast and brain / Grow to some Southern tree, / And strange-eyed constellations reign / His stars eternally."

In *Authority*, my main character, Control, hung-over and definitely not in control and caught in the grip of horrors, stares from the window of a café, bleary-eyed, at the liquor store that in his youth had been a department store and reflects.

"Long before the town of Hedley was built, there had been an indigenous settlement here, along the river and the remains of that lay beneath the liquor store. Down below the store, too, a labyrinth of limestone cradling the aquifer, narrow caves and blind albino crawfish and soft-glowing crickets. Surrounded by the crushed remains of so many creatures, loamed into the rock and soil, pushed down by the foundations of the buildings. Would that be the biologist's understanding of the street—what she would see? Perhaps she would see, too, one possible future of that space, the liquor store crumbling under an onslaught of vines and weather damage, becoming akin to the sunken, moss-covered hills near Area X."

In the Anthropocene, hauntings and similar manifestations become emissaries or transition points between the human sense of time and the geological sense of time, "Earth magnitude" as Morton puts it. In a very real sense, the weird can convey a breadth and depth that otherwise belongs to that land of seismic activity inside of a geologist's brain.

"Yeah, this place is haunted," Giant Sand sings, "but only by a ghost."

The things that haunt us in this age are often the things we care about or have some connection to, no matter how slight, and if they are also the things that *matter* we either need to become cynics or hedonists and change the things we care about so we don't care when they're destroyed, so the hauntings cannot affect us...or, more bravely and with more effort, *let* them haunt us even if it is painful, and through that haunting find some kind of act or sense of the truth that is meaningful. No matter how large. No matter how small. All while the hyperobject I am trying to pin down looms over me and shines through me and is all places and in all ways is shining out and looming over.

*Examples abound locally. Governor Rick Scott's psychopathy treated as a localized *manifestation*, once independent, now controlled by a hyperobject wraith generated by the Gulf Oil Spill. Governor Rick Scott's Department of Environmental Protection re-envisioned as a haunting transformed under the skin by malignant storytelling and infiltrating Florida—an invisible pollution of hauntings released into the world like the natural gas leak that went untreated near Porter Ranch outside of Los Angeles.

Author's note: In exploring these ideas, I may be retracing some of Timothy Morton's own ideas as expressed in interviews and essays. For the moment, I've avoided reading the bulk of this possible influence, while planning to come back to it, so as to first emphasize my own perspective as not a philosopher but a writer of weird fiction. Some portions I gave as a lecture at Vanderbilt University earlier this year. Some sentences are reworked from related essays I wrote for Electric Literature, but are here redeployed in the service of different concepts. This essay constitutes an excerpt from a nonfiction book in progress.

EDITOR BIOGRAPHIES

Torie Bosch is the editor of *Future Tense,* a partnership of *Slate,* New America, and Arizona State University that looks at the implications of emerging technologies.

Roy Scranton is the author of *Learning to Die in the Anthropocene* and the novel *War Porn.*

CONTRIBUTOR BIOGRAPHIES

Janna Avner is a creative technologist living in Los Angeles who recently co-created Femmebit, a yearly digital new media festival celebrating women artists. Janna graduated from Yale in 2012, and is currently a gallery director who curates shows, exhibits paintings, and writes as much as time permits.

Sarah Aziza has written for *Harper's, Slate, The Nation, The New Republic, The Village Voice,* and *The Rumpus,* among others. Her work focuses on issues related to the Middle East, human rights, feminism, and mental health. Outside of writing, she is also involved in social justice activism and the arts. She divides her time between New York City, the Midwest, and the Arab world. She's on Twitter as @SarahAziza1, and more of her work can be found at www.sarahaziza.com.

David Biello is the author of *The Unnatural World: The Race to Remake Civilization in Earth's Newest Age*. He is an award-winning journalist who has been reporting on the environment and energy since 1999—long enough to be cynical but not long enough to be depressed. He is the science curator for TED as well as a contributing editor at *Scientific American*. He has also written for publications ranging from *Aeon* and *Foreign Policy* to *The New York Times* and *The New Republic*. Biello has been a guest on numerous television and radio shows, and he hosts the documentary series *Beyond the Light Switch* as well as *The Ethanol Effect* for PBS. He received a BA in English from Wesleyan University and a MS in journalism from Columbia University. He currently lives with his wife, daughter, and son near a Superfund site in Brooklyn.

Brooke Borel is a journalist, editor, and author based in Brooklyn, New York. She's written on everything from cannabis pesticides to the history of fake news, but she specializes in stories about how we use technology to shape our environment. Her work has appeared in *Popular Science, The Guardian, The Atlantic, BuzzFeed News, FiveThirtyEight, Quanta Magazine,* and *Undark Magazine,* among others. Both the Alicia Patterson Foundation and the Alfred P. Sloan Foundation have funded her projects. She teaches writing at New York University and the Brooklyn Brainery. And her books are the critically acclaimed *Infested: How the Bed Bug Infiltrated Our Bedrooms and Took Over the World* and *The Chicago Guide to Fact-Checking*, which *Library Journal* named a best reference book of 2016.

Maurice Chammah is a journalist and currently a staff writer at the Marshall Project, a nonprofit newsroom in New York that reports on the criminal justice system. His writing on prisons, courts, and police has been published by *The Atlantic, The New York Times, Esquire,* and *Slate,* among others. He is at work on a book about the death penalty in Texas, his home state. More of his work is at www.mauricechammah.com.

Michael W. Clune's *White Out: The Secret Life of Heroin* was named a Best Book of 2013 by *The New Yorker*, NPR, and other venues. His most recent book is *Gamelife: A Memoir*. He's currently a professor of English at Case Western Reserve University.

Malkia A. Cyril is founder and executive director of the Center for Media Justice (CMJ) and cofounder of the Media Action Grassroots

Network. For decades, Malkia has built the capacity of racial and economic justice movements to win media rights, access, and power in a digital age. Cyril has most recently appeared in Ava DuVernay's Academy Award-nominated documentary, *13th*, released by Netflix in October 2016. Born of parents in the Black Panther Party, Cyril is now a communications strategist, a spokesperson in the fight for digital rights and freedom, a published writer, and a proud member of the Black Lives Matter Network.

Daniel Engber is a freelance journalist and contributor to *The New York Times Magazine, Radiolab, Slate, This American Life*, and *Wired*, among other outlets.

Erica Gies is an independent journalist who writes about the core requirements for life—water and energy—as well as science, critters, waste, and more. She lives in Victoria, British Columbia, and San Francisco. Her work appears in *Scientific American, Nature, The New York Times, The Economist, The Guardian*, and other publications.

Sheyna Gifford, MD, MSc, MA is a physician, science writer, and simulated astronaut for long-duration space missions. She was the health and safety officer on the longest space simulation in NASA history: the HI-SEAS IV one-year "Mars" mission. Her writing has been published in numerous newspapers, magazines, and podcasts, including *Aeon, Narratively, Nautilus*, and *StarTalk Live!* She has contributed to the creation of science documentaries for *National Geographic, Vice*, the BBC, and the History Channel. She is a captain in the civil air patrol, a STEM mentor for young women across the country, and a physician at Washington University in St. Louis, where she lives with her husband, Ben, and two cats, Trogdor and Khoshekh.

A. M. Gittlitz is a writer and bike courier living in Brooklyn. His work focuses on counterculture and radical politics of the left and right. He is the author of several zines, including *Ruin Value*, about backpacking through Europe's dying leftist infrastructure, and *KAMIKAZA*, a biography of the late Yugoslavian chaos punk Satan Panonski. Both are available from Booklyn. He is currently researching the UFOlogy-obsessed communist sect of Posadism and *Star Trek*'s socialist vision of the future. For more of A.M.'s essays visit GITTLITZ.wordpress.com.

George Johnson has been writing about science for *The New York Times* for more than 25 years. He is a two-time winner of the AAAS Science Journalism Award and the author of nine books, including *Strange Beauty: Murray Gell-Mann and the Revolution in Twentieth-Century Particle Physics, The Ten Most Beautiful Experiments, Fire in the Mind: Science, Faith and the Search for Order*, and, most recently, *The Cancer Chronicles: Unlocking Medicine's Deepest Mystery*. His books have been translated into more than 15 languages, and three of them were short-listed for the Royal Society Book Prize.

Elizabeth Kolbert is a staff writer for *The New Yorker* and the author of *The Sixth Extinction: An Unnatural History*, which won the 2015 Pulitzer Prize for general nonfiction. She is also the author of *Field Notes from a Catastrophe: Man, Nature, and Climate Change*, which grew out of a three-part series titled "The Climate of Man." Kolbert is a two-time National Magazine Award winner, and has received a Heinz Award, a Lannan Literary Fellowship, and the Rose-Walters Prize. She lives in Williamstown, Massachusetts.

Shoshana Kordova is a writer and editor focusing on health, science, finance, gender, and parenting. She is a former editor and language columnist for the Israeli daily *Haaretz* and has written for publications including *Smithsonian, Prevention, Quartz*, the *New York Times*'s parenting blog *Motherlode*, and *The Daily Beast*. She is also the cofounder of Have Faith, Will Parent (www.havefaithwill-parent.com), a virtual meeting place for parents of all religions. Shoshana has four daughters, edits books aimed at inspiring girls to do science, and spends way too much time dreaming up feminist fairy tales.

Bill McKibben is a Schumann Distinguished Scholar in Environmental Studies at Middlebury College and the founder of the global grassroots climate campaign 350.org. He wrote the first book for a general audience about climate change, *The End of Nature*, in 1989. He has 18 honorary degrees from American colleges and universities and has been arrested seven times in civil disobedience actions.

Hal Niedzviecki is a writer, speaker, culture commentator, and editor whose work challenges preconceptions and confronts readers with the offenses of everyday life. He is the author of 11 books of fiction and nonfiction and the publisher / founder of *Broken Pencil*: the magazine of zine culture and the independent arts.

Laurie Penny is a writer and journalist from London. She has reported from the front lines of social justice movements around the world, and was a 2015 Nieman Fellow at Harvard University. She is the author of six books, most recently *Bitch Doctrine: Essays for Dissenting Adults*.

Mallory Pickett is a freelance journalist with a background in analytical and marine chemistry. She left the lab in 2014 to pursue journalism full-time, and since then has written for *The New York Times Magazine*, *Wired*, *FiveThirtyEight*, and many other publications. She earned her masters in chemistry from UC San Diego and Scripps Institution of Oceanography, and her masters in journalism from UC Berkeley. She is currently based in Los Angeles.

Kim Stanley Robinson is an American science fiction writer. He is the author of more than twenty books, including the internationally bestselling *Mars* trilogy, and more recently *New York 2140*, *Aurora*, *Shaman*, *Green Earth*, and *2312*, which was a *New York Times* bestseller, nominated for all seven of the major science fiction awards—a first for any book. His work has been translated into 25 languages, and has won a dozen awards in five countries, including the Hugo, Nebula, Locus, and World Fantasy Awards. In 2016 he was given the Heinlein Award for lifetime achievement in science fiction, and asteroid 72432 was named Kimrobinson in his honor.

Jeff VanderMeer has been called "the weird Thoreau" by *The New Yorker* for his engagement with ecological issues. His most recent novel, *Borne,* received widespread critical acclaim for its exploration of animal and human life in a post-scarcity landscape. VanderMeer's prior work includes the Southern Reach trilogy (*Annihilation, Authority,* and *Acceptance*), which explored the limits of ecology and human understanding. His nonfiction has appeared in *The New York Times, The Los Angeles Times, The Atlantic, Slate, Salon,* and *The Washington Post,* among others. VanderMeer has spoken on both global warming issues and social media at MIT, Vanderbilt, DePaul, the University of Florida, the Arthur C. Clarke Center for the Human Imagination, and many others. He served as the 2016–17 Trias Writer in Residence at Hobart and William Smith Colleges.

Geoff Watts spent five years in academic biomedical research, realized he'd made a mistake in thinking he'd enjoy lab work, and dropped out with no plans for the future beyond staying in touch with science. Journalism eventually offered the ideal escape route, and he's since divided his time between writing and radio broadcasting. He's presented countless programs on science and medicine for BBC Radio 3, Radio 4, and the World Service—but has not yet learned to like the sound of his own voice.